Simplified Design of Data Converters

The EDN Series for Design Engineers

Simplified Design of Data Converters

John D. Lenk

Newnes

Boston   Oxford   Johannesburg   Melbourne   New Delhi   Singapore

Newnes is an imprint of Butterworth–Heinemann.

Copyright © 1997 by Butterworth–Heinemann

A member of the Reed Elsevier group

Recognizing the importance of preserving what has been written, Butterworth–Heinemann prints its books on acid-free paper whenever possible.

Library of Congress Cataloging-in-Publication Data

Lenk, John D.
  Simplified design of data converters / by John D. Lenk.
      p.   cm. — (EDN series for design engineers)
  Includes index.
  ISBN 0–7506–9509–9 (paper)
    1. Analog-to-digital converters—Design and construction.
  2. Digital-to-analog converters—Design and construction.
  3. Electronic circuit design.   I. Title.   II. Series.
  TK7887.6.L46   1997
  621.39′814—dc21   96-48397
                                                        CIP

British Library Cataloguing-in-Publication Data

A catalogue record for this book is available from the British Library.

The publisher offers special discounts on bulk orders of this book.
For information, please write:

Manager of Special Sales
Butterworth–Heinemann
313 Washington Street
Newton, MA 02158–1626
Tel: 617-928-2500
Fax: 617-928-2620

For information on all Newnes electronics publications available, contact our World Wide Web home page at: http://www.bh.com

10 9 8 7 6 5 4 3 2 1

Printed in the United States of America

Greetings from the Villa Buttercup!

To my wonderful wife, Irene: Thank you for being by my side all these years! To my lovely family: Karen, Tom, Brandon, Justin, and Michael. And to our Lambie and Suzzie: Be happy wherever you are! And to my special readers: May good fortune find your doorway, bringing you good health and happy things. Thank you for buying my books!

To Karen Speerstra, Jo Gilmore, Duncan Enright, Philip Shaw, Elizabeth McCarthy, the Newnes people, the UK people, and the EDN people: A special thanks for making me an international best-seller, again (this is book 87).

Abundance!

Contents

Preface

This book has something for everyone involved in electronics. No matter what your skill level, this book shows you how to design and experiment with data converters, both analog-to-digital and digital-to-analog.

For experimenters, students, and serious hobbyists, the book provides sufficient information to design and build data-converter circuits from scratch. The design approach here is the same one used in all my books on simplified and practical design.

The first three chapters provide the basics for all phases of practical design, including *testing and troubleshooting of completed circuits*. The other seven chapters include worked-out design examples that can be put to immediate use.

Throughout the book, design problems start with guidelines for selecting all components on a trial-value basis. The assumption is that one has a specific design goal and set of conditions. Then with the guideline values in experimental circuits, one can produce the desired results (resolution, accuracy, linearity, conversion rate, monotonicity) by varying the experimental component values, if needed.

If you are a working engineer responsible for designing data-converter circuits or selecting integrated-circuit data converters, the variety of circuit configurations described herein should generally simplify your task. The book not only describes converter-circuit designs but also covers the most popular forms of data-converter integrated circuits available. Throughout the book, you will find a wealth of information on data-converter integrated circuits and related components.

Chapter 1 is devoted to basic data converters, particularly circuits found in integrated-circuit form. This information is included for those who are not completely familiar with data converters and for those who need a quick refresher. The descriptions here form the basis for understanding the operation of the many integrated circuits covered in Chapters 2 through 10. Such an understanding is essential for simplified, practical design.

Chapter 2 describes data-converter terms, particularly those found on converter integrated-circuit data sheets. The emphasis is on how the listed parameters relate to design problems.

Chapter 3 concentrates on practical design considerations. Emphasis is on how the reader can select a converter to suit a specific system need but still keep over-specification (with the usual high cost) to a minimum.

Chapter 4 describes simplified-design approaches for a typical analog-to-digital integrated-circuit converter. The chapter concludes with design for a fully isolated 12-bit ADC using serial-to-parallel conversion.

Chapter 5 describes simplified-design approaches for a typical high-speed flash-type converter. The chapter concludes with design for a high-speed interface circuit.

Chapter 6 describes simplified-design approaches for a typical digital-to-analog converter, with three-wire serial interface. The chapter concludes with design for a typical bipolar output circuit.

Chapter 7 describes simplified-design approaches for a typical digital-to-analog converter with parallel input. The chapter concludes with a typical four-quadrant multiplication circuit.

Chapter 8 is devoted to simplified design approaches for a cross section of data-converter integrated circuits. The circuits in this chapter represent both classic and state-of-the-art applications.

Chapter 9 describes a converter integrated circuit that contains all major components of a data-acquisition system. Because the device operates from a single 5-V supply and draws very low power, the system is well suited to portable applications.

Chapter 10 describes a converter integrated circuit that contains the major components of a 3 ¾-digit digital multimeter and can be used for portable data-acquisition systems under microprocessor control.

Acknowledgments

Many professionals have contributed to this book. I gratefully acknowledge their tremendous effort in making this work so comprehensive: it is an impossible job for one person. I thank all who contributed, directly or indirectly.

I give special thanks to Syd Coppersmith of Dallas Semiconductor, Rosie Hinojosa of EXAR Corporation, Jeff Salter of GEC Plessey, Linda daCosta and John Allen of Harris Semiconductor, Ron Denchfield of Linear Technology, David Fullagar and William Levin of Maxim Integrated Products, Fred Swymer of Microsemi Corporation, Linda Capcana of Motorola, Inc., Andrew Jenkins and Shantha Natrajan of National Semiconductor, Antonio Ortiz of Optical Electronics, Inc., Lawrence Fogel of Philips Semiconductors, John Marlow of Raytheon Company Semiconductor Division, Anthony Armstrong of Semtech Corporation, Ed Oxner and Robert Decker of Siliconix Inc., Amy Sullivan of Texas Instruments, and Diane Freed Publishing Services.

I also thank Joseph A. Labok of Los Angeles Valley College for help and encouragement throughout the years.

Very special thanks to Karen Speerstra, Jo Gilmore, Duncan Enright, Philip Shaw, Elizabeth McCarthy, Joan Dargan, Drew Bourn, Marika Alzadon, Karen Burdick, Dawn Doucette, Laurie Hamilton, the Newnes people, the UK people, and the EDN people of Butterworth–Heinemann for having so much confidence in me. I recognize that all books are a team effort and am thankful that I now work with the New First Team on this series.

And to Irene, my wife and super agent, I extend my thanks. Without her help, this book could not have been written.

CHAPTER 1

Data Converter Basics

This chapter is devoted to basic data converters, both analog-to-digital and digital-to-analog. The chapter is primarily for readers who are totally unfamiliar with data converters. It is possible to design data-converter circuits from scratch. However, data converters are available in integrated circuit (IC) form, and it is generally simpler to use such ICs.

The data sheets for IC converters often show the connections and provide all necessary design parameters to produce complete converter circuits by adding external components. This chapter describes the function and operation of IC converters to help you understand the data sheet information.

Before we get started, let us resolve certain differences in terms. Some manufacturers refer to analog-to-digital converters as *ADCs*. Other manufactures use the term *A/D converter.* The same is true of digital-to-analog converters, which are referred to as *DACs* by some, and as *D/A converters* by others. I prefer the terms *ADC* and *DAC,* but do not be surprised to find both terms in this book.

1.1 Basic Data Conversion Techniques

This section describes the various ADC and DAC techniques in common use. Here we concentrate on explanations of the basic principles of data conversion. By studying this information, you should be able to understand operation of the converter IC described throughout the book. It is assumed that you are familiar with basic digital electronics. If not, read Lenk's *Digital Handbook* (McGraw-Hill, 1993).

1.1.1 Typical BCD Signal Formats Used in ADC/DAC Circuits

Figure 1-1a shows the relationship of the three most common BCD (binary coded decimal) signal formats: NRZL (nonreturn-to-zero level), NRZM (nonreturn-to-zero-mark), and RZ (return-to-zero).

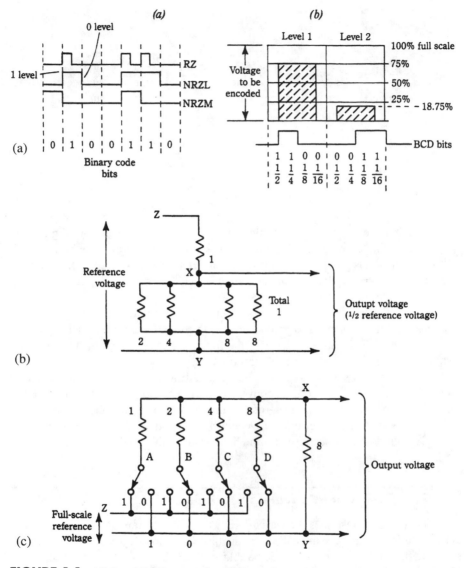

FIGURE 1-1 ADC and DAC conversion basics

In NRZL, a 1 is one signal level, and a 0 is another signal level. These levels can be 5 V, 10 V, 0 V, −5 V, or any other selected values, provided that the 1 and 0 levels are entirely different.

In RZ, a 1-bit is represented by a pulse of definite width (usually a ½-bit width) that returns to zero signal level, and the 0-bit is represented by a zero-level signal.

In NRZM, the level of the pulse has no meaning. A 1 is represented by a change in level, and a 0 is represented by no change.

1.1.2 Four-Bit System in the Conversion Process

Figure 1-1b shows the relation between two voltage levels to be converted, and the corresponding binary code (in NRZL form), in a basic ADC. In practice, a four-bit ADC (sometimes called a *binary encoder*) samples the voltage level to be converted and compares the voltage to ½ scale, ¼ scale, ⅛ scale, and ¹⁄₁₆ scale (in that order) of a given full-scale voltage. The ADC then produces four data bits, in sequence, with the comparison made on the most significant (½ scale) first.

As shown in Fig. 1-1b, each of the two voltage levels is divided into four equal time increments. The first time increment is used to represent the ½-scale bit, the second increments the ¼-scale, and so on.

In voltage level 1, the first two time increments are at binary 1, and the second two increments are at 0. This produces 1100, or decimal 12. Twelve is ¾ of 16. Thus level 1 is 75% of full scale. For example, if full scale is 10 V, level 1 is 7.5 V.

In level 2, the first two increments are at 0, and the second two increments are at 1. This is represented as 0011, or 3. Thus level 2 is ³⁄₁₆ of full scale (or 1.875 V). This can be expressed in another way. In the first or ½-scale increment, the converter produces a 0 because the voltage (1.875 V) is less than ½ scale (5 V). The same is true of the second or ¼-scale increment (1.875 V is less than 2.5 V).

In the third or ⅛-scale increment of level 2, the converter produces a 1, as it does in the fourth or ¹⁄₁₆-scale increment, because the voltage being compared is greater than ⅛ of full scale (1.875 is greater than 0.625 V). Thus the ½- and ¼-scale increments are at 0, and the ⅛- and ¹⁄₁₆-scale increments are at 1 (also, ⅛ + ¹⁄₁₆ = ³⁄₁₆ or 18.75%).

1.1.3 ADC Conversion Ladder

Figure 1-1c shows a conversion ladder, which is the heart of many ADC circuits. The ladder provides a means of implementing a four-bit binary-coding system and produces an output that is equivalent to switch positions. The switches can be moved to either a 1 or a 0 position, which corresponds to a four-place binary number. The output voltage describes a percentage of the full-scale reference voltage, depending on the switch positions. For example, if all switches are at 0 position, there is no output voltage. This produces a binary 0000, represented by 0 V.

If switch A is at 1 and the remaining switches are at 0, this produces a binary 1000 (decimal 8). Because the total in a four-bit system is 16 (0 to 15), 8 represents ½ full scale. Thus the output voltage is ½ the full-scale reference voltage. This conversion is done as follows.

The 2-, 4-, and 8-ohm switch resistors and the 8-ohm output resistor are connected in parallel. This produces a value of 1 ohm across points X and Y. The reference voltage is applied across the 1-ohm switch resistor (across points Z and X) and the 1-ohm combination of resistors (across points X and Y); in effect, this is the same as two 1-ohm resistors in series. Because the full-scale reference voltage is applied across both resistors in series, and the output is measured across only one of the resistors, the output voltage is ½ of the reference voltage.

In a practical converter, the same basic ladder is used to supply a comparison voltage to a comparison circuit, which compares the voltage to be converted against the binary-coded voltage from the ladder. The resultant output of the comparison circuit is a binary code representing the voltages to be converted.

The mechanical switches shown in Fig. 1-1c are replaced by electronic switches, usually flip-flops (FFs). When the switch is on, the corresponding ladder resistor is connected to the reference voltage. The switches are triggered by four pulses (representing each of the four binary bits) from the clock. An enable pulse is used to turn the comparison circuit on and off, so that as each switch is operated, a comparison can be made of the four bits.

1.1.4 Typical ADC Operating Sequence

Figure 1-2a is a simplified diagram of a typical ADC. Here the reference voltage is applied to the ladder through the electronic switches. The ladder output (comparison voltage) is controlled by switch positions, which are controlled by pulses from the clock.

The following paragraphs outline the sequence of events necessary to produce a series of four binary bits that describe the input voltage as a percentage of full scale (in $\frac{1}{16}$ increments). Assume that the input voltage is 75% of full scale.

When pulse 1 arrives, switch 1 is turned on and the remaining switches are off. The ladder output is a 50% voltage that is applied to the differential amplifier. The balance of this amplifier is set so that the output is sufficient to turn on one AND gate and turn off the other AND gate, if the ladder voltage is greater than the input voltage. Similarly, the differential amplifier reverses the AND gates if the ladder voltage is not greater than the input voltage. Both AND gates are enabled by the clock pulse.

In this example (75% of full scale), the ladder output is less than the input voltage when pulse 1 is applied to the ladder. As a result, the not-greater AND gate turns on, and the output FF is set to the 1 position. Thus for the first of the four bits, the FF output is 1.

When pulse 2 arrives, switch 2 is turned on, and switch 1 remains on. Both switches 3 and 4 remain off. The ladder output is now 75% of the full-scale voltage. (The ladder voltage equals the input voltage.) However, the ladder output is still not greater than the input voltage. Consequently, when the AND gates are enabled, the AND gates remain in the same condition. Thus the output FF remains at 1.

When pulse 3 arrives, switch 3 is turned on. Switches 1 and 2 remain on with switch 4 off. The ladder output is now 87.5% of full-scale voltage, and thus is greater than the input voltage. As a result, when the AND gates are enabled, they reverse. The not-greater AND gate turns off, and the greater AND gate turns on. The output FF then sets to 0.

When pulse 4 arrives, switch 4 is turned on. All switches are now on. The ladder is at maximum (full scale) and this is greater than the input voltage. As a result, when the AND gates are enabled, they remain in the same condition. The output FF remains at 0.

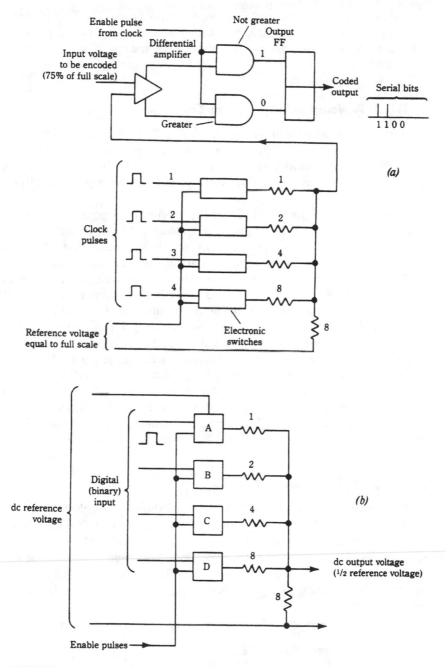

FIGURE 1-2 ADC and DAC conversion circuits

The four binary bits from the output are 1, 1, 0, and 0, or 1100. This is a binary 12, which is 75% of 16. In a practical ADC, when the fourth pulse has passed, all switches are reset to the off position. This places them in a condition to spell out the next four-bit binary word.

1.1.5 Typical DAC Operating Sequence

Figure 1-2b is a simplified diagram of a typical DAC. This circuit performs the opposite function of the ADC just described (the DAC produces an output voltage that corresponds to the binary code). A conversion ladder is also used in the DAC. The conversion-ladder output is a voltage that represents a percentage of the full-scale reference voltage.

The output voltage from the DAC also depends on switch positions. In turn, the switches are set to on or off by corresponding binary pulses. If the information is applied to the switch in four-line (parallel) form, each line can be connected to the corresponding switch. If the information is in serial form, the data must be converted to parallel by a register (shift or storage register). The switches in a DAC are essentially a form of AND gate. Each gate completes the circuit from the reference voltage to the corresponding ladder resistor when both the enable pulse and binary pulse coincide.

Assume that the digital number to be converted is 1000 (decimal 8). When the first pulse is applied, switch A is enabled and the reference voltage is applied to the 1-ohm resistor. When switches B, C, and D receive their enable pulses, there are no binary pulses (or the pulses are in the 0 condition). Thus switches B, C, and D do not complete the circuits to the 2-, 4-, and 8-ohm ladder resistors. These resistors combine with the 8-ohm output resistor to produce a 1-ohm resistance in series with the 1-ohm ladder resistance. This divides the reference voltage in half to produce 50% of full-scale output. Because 8 is ½ of 16, the 50% output voltage represents 8.

1.1.6 High-Speed ADCs

Although there are a number of ADC schemes, there are only three basic types: parallel, serial, and combination. In parallel, all bits are converted simultaneously by many circuits. In serial, each bit is converted in sequence, one at a time. The combination ADC includes features of both types. Generally, parallel is faster but more complex than serial. The combination types are a compromise between speed and complexity. The remaining paragraphs in this section describe some classic high-speed ADC circuits.

1.1.7 Parallel (Flash) ADC

Figure 1-3a shows the basic parallel (or flash) ADC circuit, where all bits of the digital representation are converted simultaneously by a bank of voltage comparators. For N bits of binary information, the system requires 2^{N-1} comparators, and each comparator determines one LSB (least significant bit) level. This requires a

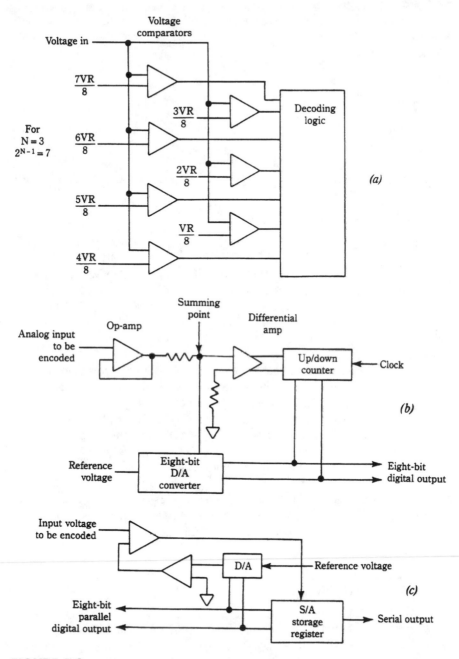

FIGURE 1-3 High-speed ADC converters

number of circuits. Another disadvantage of parallel ADC is that comparator output is not directly usable information. The output must be converted to binary information with a decoder.

1.1.8 Tracking ADC

Figure 1-3b shows the basic tracking ADC circuit. A tracking ADC continuously tracks the analog input voltage and is often used in communication systems or similar applications in which the input is a continuously varying signal. The accuracy of the system is no better than the DAC used in the system. (An 8-bit D/A converter is shown in Fig. 1-3b.)

1.1.9 Successive Approximation ADC

Figure 1-3c shows the basic successive approximation ADC circuit. Note that this circuit is essentially the same as the basic ADC of Fig. 1-2a. The D/A block of Fig. 1-3c represents the electronic switches and ladder of Fig. 1-2a. The successive approximation (S/A) storage register (often called an SAR) of Fig. 1-3c represents the AND gates and FF of Fig. 1-2a. However, four bits are shown in Fig. 1-2a, whereas eight bits are used in Fig. 1-3c.

The successive-approximation type of ADC is relatively slow compared with other types of high-speed ADCs, but the low cost, ease of construction, and system features make up for the lack of speed. With successive approximation, eight bits of the D/A are enabled, one at a time, starting with the MSB (most significant bit). As each bit is enabled, the comparator produces an output indicating that the input signal is greater, or not greater, in amplitude than the output of the D/A. If the D/A output is greater than the input signal, the bit is reset or turned off. The system does this with the MSB first, then the next most significant bit, then the next, and so on. After all eight bits of the D/A are tried, the conversion cycle is complete, and another cycle is started. Note that the S/A type of ADC provides a serial output during conversion and a parallel output between conversion cycles.

1.2 Typical DAC IC

Figure 1-4 shows the functional block diagram of a typical DAC IC (the classic DAC-08). Figure 1-5 shows the connection information and thermal characteristics for the IC. Such thermal characteristics are essential for simplified design with any IC, especially where heat sinks are involved. If you are not familiar with heat sinks and IC thermal problems, read my *Simplified Design of Linear Power Supplies* (Butterworth–Heinemann, 1994). Figure 1-6 shows the DAC-08 connected for basic operation with a positive reference voltage.

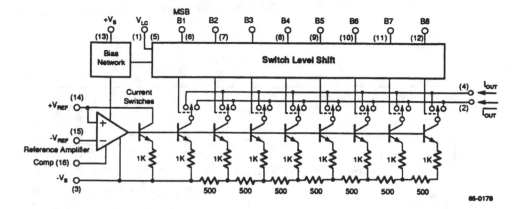

FIGURE 1-4 Functional block diagram of typical DAC IC (*Raytheon Semiconductor Data Book,* 1994, p. 3-141)

1.2.1 Basic Design Requirements

The DAC-08 is a *multiplying* DAC in which the output current is the product of a digital number and the input reference current. This is somewhat different from the theoretic DAC of Fig. 1-2b, but the net result is the same. A digital or binary word is converted to an output current (which can be further converted to voltage by passing the current through a resistor or load).

In the circuit of Fig. 1-6, the reference current may be fixed or vary from 100 uA to 4 mA. The full-scale output current (IFS) is a linear function of the reference current (IREF) and is given by IFS = (255/256) × IREF, where IREF = I14. In the positive-reference application shown, the external reference forces current through R14 into the +V_REF terminal (pin 14) of the reference amplifier (Fig. 1-4). If required, a negative reference may be applied to the −V_REF terminal (pin 15). This negative-reference connection has the advantage of a very high impedance presented at pin 15. The voltage at pin 14 is equal to and tracks the voltage at pin 15. In some cases, R15 can be eliminated with only minor increases in tracking error. R15 is used to cancel any input bias current errors in the reference amplifier. Note that the reference amplifier is essentially an op amp (operational amplifier) and requires all of related design considerations (input bias current, phase margin correction). If you are not familiar with op-amp design problems, read my *Simplified Design of IC Amplifiers* (Newnes, 1996).

Figure 1-7 shows how bipolar references can be accommodated. Either V_REF or pin 15 can be offset. The negative common-mode range of the reference amplifier is given by VCM = −VS plus (IREF × 1k) plus 2.5 V. The positive common-mode range is +VS less 1.5 V. When the reference is DC, a reference bypass capacitor is recommended. If a regulated power supply is used as a reference, R14 should be split into two resistors with the junction bypassed to ground with a 0.1-μF capacitor. (A 5-V TTL-logic supply is not recommended as a reference.)

Thermal Characteristics

	16-Lead Ceramic DIP	16-Lead Plastic DIP
Max. Junction Temp.	+175°C	+125°C
Max. P_D T_A <50°C	1042 mW	555 mW
Therm. Res θ_{JC}	60°C/W	—
Therm. Res. θ_{JA}	120°C/W	135°C/W
For T_A >50°C Derate at	8.38 mW/°C	7.41 mW/°C

Connection Information

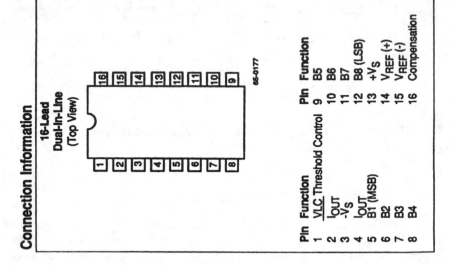

16-Lead Dual-In-Line (Top View)

65-0177

Pin	Function	Pin	Function
1	VLC Threshold Control	9	B5
2	I_{OUT}	10	B6
3	$-V_S$	11	B7
4	$\overline{I_{OUT}}$	12	B8 (LSB)
5	B1 (MSB)	13	$+V_S$
6	B2	14	V_{REF} (+)
7	B3	15	V_{REF} (-)
8	B4	16	Compensation

FIGURE 1-5 Connection information and thermal characteristics of typical DAC IC (*Raytheon Semiconductor Data Book*, 1994, p. 3-140)

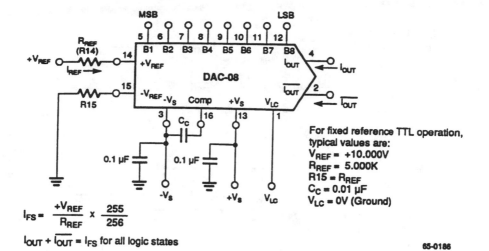

FIGURE 1-6 DAC-08 connected for basic positive-reference operation (*Raytheon Semiconductor Data Book,* 1994, p. 3-148)

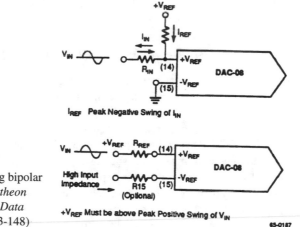

FIGURE 1-7

Accommodating bipolar references (*Raytheon Semiconductor Data Book,* 1994, p. 3-148)

Figure 1-8 shows how the full-scale circuit can be adjusted. In most cases, this is not necessary because of the close relationship between IREF and IFS. It is also possible to adjust full-scale by substituting a potentiometer for R14. However, this is not recommended because of the temperature-coefficient (TC) effects of a potentiometer.

Figure 1-9 shows the basic negative reference operation. Using lower values of reference current reduces negative power-supply current and increases reference-amplifier negative common-mode range. The recommended range for operation with

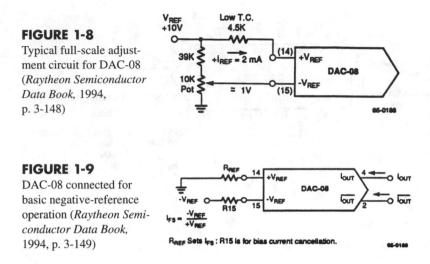

FIGURE 1-8

Typical full-scale adjustment circuit for DAC-08 (*Raytheon Semiconductor Data Book,* 1994, p. 3-148)

FIGURE 1-9

DAC-08 connected for basic negative-reference operation (*Raytheon Semiconductor Data Book,* 1994, p. 3-149)

a DC reference current is +0.2 mA to +4.0 mA. With either positive or negative reference operation, the reference amplifier must be compensated by a capacitor C_c connected from pin 16 to −VS, as shown in Fig. 1-6. If the reference is a fixed DC voltage, a value of 0.01 μF is recommended for the compensating capacitor. If the reference is AC or pulse, the value of C_c must be selected as described in Section 1.2.2.

1.2.2 Reference Amplifier Compensation

The value of capacitor C_c depends on the impedance presented to pin 14. For example, for R14 values of 1.0, 2.5, and 5.0 k, the minimum values of C_c are 15, 37, and 75 pF, when the reference is AC. Larger values of R14 require proportionately increased values of C_c to ensure proper phase margin. When the reference input is a pulse, use a low value for R14. This makes it possible to use smaller values for C_c.

If pin 14 is driven by a high impedance, such as a transistor current source, the recommended values for compensation do not apply. This is because the reference amplifier must be heavily compensated. Unfortunately, such compensation decreases overall bandwidth and slew rate (as is the case when an op amp is heavily compensated). As a point of reference, for an R14 of 1 k and a C_c of 15 pF, the reference amplifier slews at 4.0 mA/μs, and results in a transition from IREF = 0 to IREF = 2.0 mA in 500 ns.

When a low full-scale transition time is critical, use 200 ohms for R14, and make C_c = 0. Under these conditions, full-scale transition (0 to 2.0 mA) occurs in 120 ns. This produces a slew rate of 16 mA/μs.

1.2.3 Input Circuits

The DAC-08 can be interfaced directly to all popular logic families with a maximum of noise immunity. This is because of the large input swing capability, a

2.0-μA logic-input current, and adjustable logic-threshold voltage. For example, when the supply is −15 V, the logic inputs can swing between −10 V and +18 V. This allows direct interface with +5 V CMOS (complementary metal oxide semi-conductor) logic, even when the supply is +5 V. The minimum input-logic swing and minimum logic threshold are given by: supply plus (IREF × 1 k) plus 2.5 V.

The logic threshold can be adjusted over a wide range by placing an appropriate voltage at the logic threshold control VLC (pin 1). The logic-threshold voltage VTH is nominally 1.4 V above VLC. For TTL and DTL interface, simply ground pin 1 as shown in Fig. 1-6. For interfacing ECL, an IREF of 1 mA is recommended. Figure 1-10 shows typical interface circuits for ECL (emitter-coupled logic), CMOS, and PMOS/NMOS.

Note that pin 1 will source or sink 100 μA (typical), so external circuits should be designed to accommodate this current. When a fast settling time (Section 1.2.7) is essential, keep in mind that the fastest times are obtained when pin 1 sees a low impedance. For example, pin 1 can be connected to a 1-k divider and bypassed to ground with a 0.01 μF capacitor.

1.2.4 Output Circuits

Figures 1-11 through 1-13 show typical unipolar, bipolar, and offset-binary output circuits and the relationships between digital inputs and voltage/current outputs. As shown, both true and complemented output sink currents are provided where $IOUT + \overline{IOUT} = IFS$.

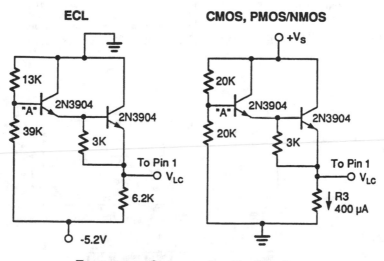

FIGURE 1-10 Typical interface circuits for ECL, CMOS, and PMOS/NMOS (*Raytheon Semiconductor Data Book,* 1994, p. 3-150)

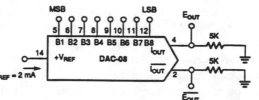

FIGURE 1-11

DAC-08 connected for basic unipolar-negative operation (*Raytheon Semiconductor Data Book,* 1994, p. 3-149)

Scale	B1	B2	B3	B4	B5	B6	B7	B8	I$_{OUT}$mA	Ī$_{OUT}$mA	E$_{OUT}$	Ē$_{OUT}$
Full Scale	1	1	1	1	1	1	1	1	1.992	0.008	-9.660	-0.000
Half Scale +LSB	1	0	0	0	0	0	0	1	1.008	0.984	-5.040	-4.920
Half Scale	1	0	0	0	0	0	0	0	1.000	0.992	-5.000	-4.960
Half Scale -LSB	0	1	1	1	1	1	1	1	0.992	1.000	-4.960	-5.000
Zero Scale +LSB	0	0	0	0	0	0	0	1	0.008	1.984	-0.040	-9.920
Zero Scale	0	0	0	0	0	0	0	0	0.000	1.992	0.000	-9.960

65-0190

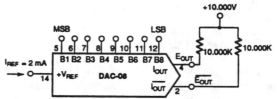

FIGURE 1-12

DAC-08 connected for basic bipolar-output operation (*Raytheon Semiconductor Data Book,* 1994, p. 3-149)

Scale	B1	B2	B3	B4	B5	B6	B7	B8	E$_{OUT}$	Ē$_{OUT}$
Pos Full Scale	1	1	1	1	1	1	1	1	-9.920	+10.000
Pos Full Scale - LSB	1	1	1	1	1	1	1	0	-9.840	+9.920
Zero Scale + LSB	1	0	0	0	0	0	0	1	-0.080	+0.160
Zero Scale	1	0	0	0	0	0	0	0	0.000	+0.080
Zero Scale - LSB	0	1	1	1	1	1	1	1	+0.080	0.000
Neg Full Scale + LSB	0	0	0	0	0	0	0	1	+9.920	-9.840
Neg Full Scale	0	0	0	0	0	0	0	0	+10.000	-9.920

65-0191

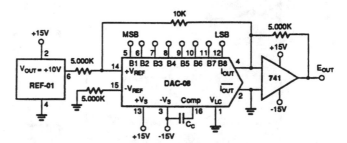

FIGURE 1-13

DAC-08 connected for basic offset-binary operation (*Raytheon Semiconductor Data Book,* 1994, p. 3-149)

Scale	B1	B2	B3	B4	B5	B6	B7	B8	E$_{OUT}$
Pos Full Scale	1	1	1	1	1	1	1	1	+4.960
Zero Scale	1	0	0	0	0	0	0	0	0.000
Neg Full Scale + 1 LSB	0	0	0	0	0	0	0	1	-4.960
Neg Full Scale	0	0	0	0	0	0	0	0	-5.000

65-0192

Current appears at the true output when a 1 is applied to each logic input. As the binary count increases, the sink current at pin 4 increases proportionally (a positive-logic DAC). When a 0 is applied to any input bit, that current is turned off at pin 4 and turned on at pin 2. A decreasing logic count increases IOUT proportionally (a negative-logic or inverted-logic DAC).

Both outputs can be used simultaneously, and both have a wide voltage compliance. This allows fast current-to-voltage conversion through a resistor tied to ground or other voltage source (Figs. 1-11, 1-12). Positive compliance is 36 V above the negative supply and is independent of the positive supply. Negative compliance is given by supply voltage plus (IREF $\times$ 1 k) plus 2.5 V.

The dual outputs allow double the usual peak-to-peak load swing when driving loads in a quasidifferential manner. This feature is especially useful in cable-driver or CRT-deflection circuits and other balanced applications, such as driving center-tapped coils and transformers.

If one of the outputs is not required, it must still be connected to ground or to a point capable of sourcing IFS. *Do not* leave an unused output pin open. This is shown in Figs. 1-14 and 1-15 where the pin-4 output is connected to an op amp. With these circuits, the output impedance is lowered to a value determined by the op amp, not the DAC. In both cases, the unused DAC output is connected to ground and the converted output is taken from the op amp.

1.2.5 Power Supply Requirements

Symmetric supplies are not required. However, the DAC-08 will operate satisfactorily with standard $\pm$5.0-V supplies, or any supply combination in which the total is 9 V to 36 V. If $\pm$5.0 V is used, the IREF should not exceed 1 mA. Keep in mind that low reference-current operation decreases power consumption and increases negative compliance, reference-amplifier negative common-mode range, negative-logic input range, and negative-logic threshold range. This is shown in the various graphs of Fig. 1-16.

Power consumption (PD) can be calculated as follows: (I+ $\times$ +VS) + (I- $\times$ -VS) + (2 IREF $\times$ -VS). Supply current (I) is constant and independent of input-logic states. This reduces the size of power-line bypass capacitors.

1.2.6 Temperature Characteristics

The nonlinearity and monotonicity specifications of the DAC-08 are guaranteed to apply over the entire rated operating temperature range (Fig. 1-5). (See Chapter 2 for a discussion of monotonicity.) Full-scale output-current drift is typically $\pm$10 ppm/°C. Zero-scale output current and drift are essentially negligible to $\frac{1}{2}$ LSB.

To achieve minimum overall full-scale drift, the temperature coefficient of the reference resistor (Fig. 1-6) should match and track that of the output resistor. Settling times (Section 1.2.7) decrease approximately 10% at -55°C. At +125°C, an increase of about 15% is typical.

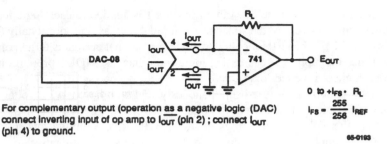

For complementary output (operation as a negative logic (DAC) connect inverting input of op amp to $\overline{I_{OUT}}$ (pin 2) ; connect I_{OUT} (pin 4) to ground.

0 to $+I_{FS} \cdot R_L$

$I_{FS} = \dfrac{255}{256} I_{REF}$

65-0193

FIGURE 1-14 DAC-08 connected for positive low-impedance output operation (*Raytheon Semiconductor Data Book,* 1994, p. 3-150).

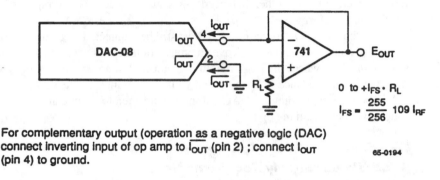

For complementary output (operation as a negative logic (DAC) connect inverting input of op amp to $\overline{I_{OUT}}$ (pin 2) ; connect I_{OUT} (pin 4) to ground.

0 to $+I_{FS} \cdot R_L$

$I_{FS} = \dfrac{255}{256} \; 109 \; I_{RF}$

65-0194

FIGURE 1-15 DAC-08 connected for negative low-impedance output operation (*Raytheon Semiconductor Data Book,* 1994, p. 3-150)

1.2.7 Settling Time

Figure 1-17 shows a settling-time test circuit for the DAC-08. This circuit also can be used for many other DACs with slight modification. Settling time is often the single most critical factor in DAC operation because it determines overall speed of the digital-to-voltage conversion. The DAC-08 requires 35 ns for each of the 8 data bits. Settling time to within ½ LSB is therefore 35 ns, with each progressively larger bit taking successively longer. The MSB settles in 85 ns, determining the overall settling time of 85 ns. Figure 1-16 shows a typical oscilloscope response for measurement of full-scale settling time, indicating about 85 ns from the start of the logic input to the point where the output settles to a constant level. Settling to 6-bit accuracy requires about 65 to 70 ns.

As in the case of all digital ICs, the fastest operation is obtained using short leads, minimum output capacitance, minimum load resistance, and adequate bypassing for the supply. The bypass need not be electrolytic because supply current drain is independent of the input logic state. The 0.1-µF bypass capacitors shown provide full transient protection. The output capacitance of the DAC-08 (including the package) is about 15 pF. Therefore the output RC (resistance-capacitance) time constant dominates settling time when the load is greater than 500 ohms.

Typical Performance Characteristics

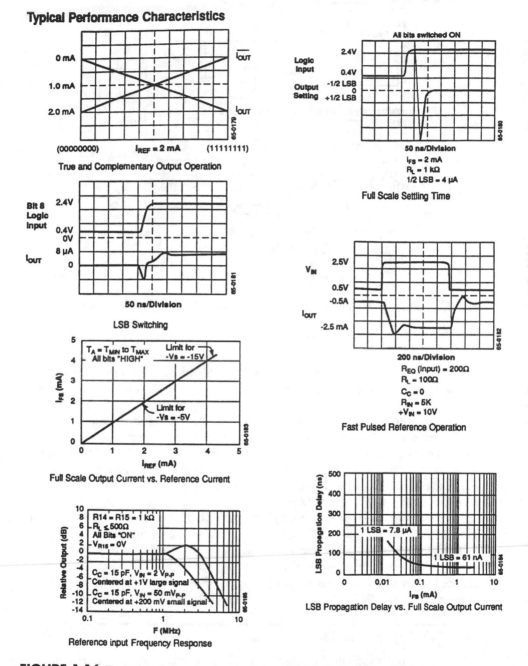

FIGURE 1-16 Typical performance characteristics of DAC-08 (*Raytheon Semiconductor Data Book,* 1994, p. 3-145)

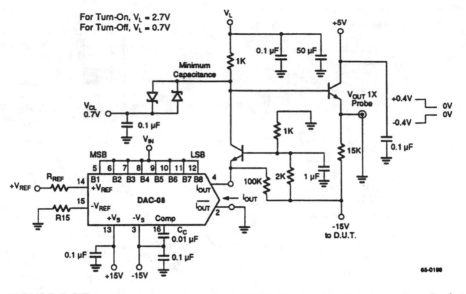

FIGURE 1-17 Settling-time test circuit for DAC-08 (*Raytheon Semiconductor Data Book*, 1994, p. 3-151)

Measurement of settling time requires the ability to accurately resolve ±4.0 μA, so a 1-k load is needed to provide adequate drive (±0.4 V) for most scopes. The circuit of Fig. 1-17 uses a cascade design to allow driving the 1-k load with less than 5 pF of parasitic capacitance at the measurement node. At IREF values of less than 1 mA, excessive RC damping of the output is difficult to prevent if adequate sensitivity is maintained. However, the major carry from 01111111 to 10000000 provides an accurate indicator of settling time. This code change does not require the normal 6.2 time constants to settle to within ±0.2% of the final value, so settling times can be observed at lower values of IREF.

Settling time remains essentially constant for IREF values down to 1 mA, with gradual increases for lower IREF values. The main advantage of higher IREF values is the ability to obtain a given output level with lower load resistors, reducing the output RC time constant. Switching transients are generally low and can be further reduced by small capacitive loads at the output. Of course, this increase in RC time constant does increase settling time.

1.3 Typical ADC IC

Figure 1-18 shows the functional block diagram of typical ADC ICs (the industry standard MAX174 and MX574A/MX674A). Figure 1-19 shows the pin functions. The ICs are complete 12-bit ADCs that combine high speed, low-power consumption, and on-chip clock and voltage reference. The maximum conversion times are 8 ms (MAX714), 15 ms (MX674A), and 25 ms (MX574A).

1.3.1 Basic Converter Operation

These ICs use the successive-approximation technique described in Section 1.1.9 to convert an unknown analog input to a 12-bit digital output code. Compare the circuit of Fig. 1-18 with that of Fig. 1-3c. The control logic function of Fig. 1-18 provides easy interface with most microprocessors. The internal voltage-output DAC

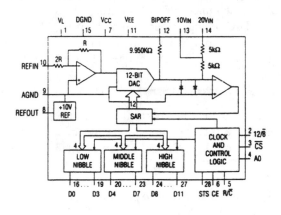

FIGURE 1-18

Functional block diagram of typical ADC IC (*Maxim New Releases Data Book,* 1992, p. 7-81)

FIGURE 1-19

Pin descriptions for typical ADC IC (*Maxim New Releases Data Book,* 1992, p. 7-87)

PIN #	NAME	FUNCTION
1	V_L	Logic Supply, +5V
2	12/$\overline{8}$	Data Mode Select Input
3	$\overline{CS}$	Chip-Select Input. Must be low to select device.
4	A0	Byte Address/Short Cycle Input. When starting a conversion, controls number of bits converted (low = 12 bits, high = 8 bits). When reading data, if 12/8 = low, enables low byte (A0 = high) or high byte (A0 = low).
5	R/$\overline{C}$	Read/Convert Input. When high, the device will be in the data-read mode. When low, the device will be in the conversion start mode.
6	CE	Chip-Enable Input. Must be high to select device.
7	V_{CC}	+12V or +15V Supply
8	REFOUT	+10V Reference Output
9	AGND	Analog Ground
10	REFIN	Reference Input
11	V_{EE}	-12V or -15V Supply
12	BIPOFF	Bipolar Offset Input. Connect to REFOUT for bipolar input range.
13	10V_{IN}	10V Span Input
14	20V_{IN}	20V Span Input
15	DGND	Digital Ground
16-27	D0-D11	Three-State Data Outputs
28	STS	Status Output

is controlled by an SAR (see Section 1.1.9). The analog input is connected to the DAC output with a 5-k resistor for the 10-V input and a 10-k resistor for the 20 V input. The comparator is essentially a zero-crossing detector, with the output fed back to the SAR input. Figure 1-20 shows the equivalent of the analog input circuit.

In the IC of Fig. 1-18, the SAR is set to half scale as soon as a conversion starts. The analog input is compared to ½ of the full-scale voltage. The bit is kept if the analog input is greater than half scale or dropped if smaller. The next bit (bit 10) is then set with the DAC output either at ¼ scale, if the MSB is dropped, or ¾ scale if the MSB is kept. The conversion continues in this manner until the LSB is tried. At the end of the conversion, the SAR output is latched into the output buffers.

1.3.2 Digital Control

Operation of the ICs is controlled by the CE (chip enable), $\overline{CS}$ (chip select), and R/$\overline{C}$ lines. Figure 1-21 shows the truth table for these lines and the A0 (pin 4) and ¹²⁄₈ (pin 2) lines. While both CE and $\overline{CS}$ are asserted, the state of R/$\overline{C}$ selects whether a conversion (R/$\overline{C}$ = 0) or a data read (R/$\overline{C}$ = 1) is in progress. The register-control inputs, 12/$\overline{8}$ and A0, select the data format and conversion length. A0 is usually tied to the LSB of the address bus. To perform a full 12-bit conversion, set A0 low during a convert start. For a shorter 8-bit conversion, A0 must be high during a convert start.

1.3.3 Output Data Formats

Figure 1-22 shows the data format for an 8-bit bus, including typical hard-wiring. Output data bits are formatted according to the control signal on the ¹²⁄₈ input. If pin 2 is low, the output is a word broken into two 8-bit bytes. If pin 2 is high, the output is one 12-bit word. During a data read, A0 selects whether the three-state buffers contain the 8 MSBs (A0 = 0) or the 4 LSBs (A0 = 1) of the digital result. The 4 LSBs are followed by four trailing 0s (zeros). A0 can change state while a data-read operation is in effect.

To begin a conversion, the microprocessor must write to the ADC address. Then, because a conversion usually takes longer than a single clock cycle, the microprocessor must wait for the ADC to complete the conversion. Valid data bits are made

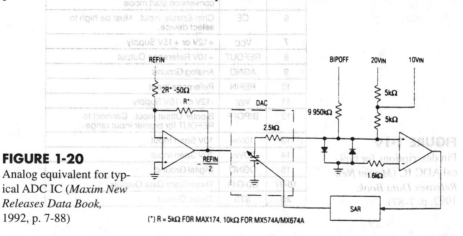

FIGURE 1-20

Analog equivalent for typical ADC IC (*Maxim New Releases Data Book,* 1992, p. 7-88)

(*) R = 5kΩ FOR MAX174, 10kΩ FOR MX574A/MX674A

CE	C̄S̄	R/C̄	12/8̄	A0	OPERATION
0	X	X	X	X	None
X	1	X	X	X	None
1	0	0	X	0	Initiate 12-bit conversion
1	0	0	X	1	Initiate 8-bit conversion
1	0	1	1	X	Enable 12-bit parallel output
1	0	1	0	0	Enable 8MSBs
1	0	1	0	1	Enable 4LSBs + 4 trailing 0s

FIGURE 1-21

Truth table for digital control of ADC IC (*Maxim New Releases Data Book,* 1992, p. 7-88)

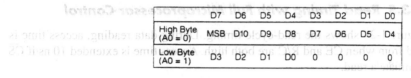

	D7	D6	D5	D4	D3	D2	D1	D0
High Byte (A0 = 0)	MSB	D10	D9	D8	D7	D6	D5	D4
Low Byte (A0 = 1)	D3	D2	D1	D0	0	0	0	0

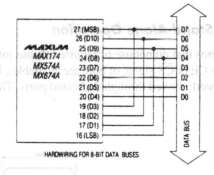

FIGURE 1-22

Data format for an 8-bit bus (*Maxim New Releases Data Book,* 1992, p. 7-89)

available only at the end of the conversion, which is indicated by STS (pin 28k the output status). STS can be either polled or used to generate an interrupt upon completion. Or the microprocessor can be kept idle by means of insertion of the appropriate number of NOP (no operation) instructions between the conversion-start and data-read commands.

After the conversion is completed, data can be obtained by the microprocessor. The ICs have the required logic for 8-, 12-, and 16-bit bus interfacing (as determined by the 12/8 input). If pin 2 is high, the ICs are configured for a 16-bit bus. Data lines D0-D11 can be connected to the bus as either the 12 MSBs or the 12 LSBs. The other 4 bits must be masked out in software.

For 8-bit bus operation, pin 2 is low. The format is left-justified and the even address (A0 low) contains the 8 MSBs. The odd address (A0 high) contains the 4 LSBs, followed by four trailing 0s. There is no need to use a software mask when the ICs are connected to an 8-bit bus. (Note that the output cannot be forced to a right-justified format by rearranging the data lines on the 8-bit bus interface.)

1.3.4 Convert-Start Timing with Full Microprocessor Control

Figure 1-23 shows the convert-start timing when the ICs are under full micro-processor control. It is essential that R/C̄ must be low before asserting both CE and C̄S. If R/C̄ is high, a brief read operation occurs. This could result in system-bus contention (a bus fight). Either CE or C̄S can be used to initiate conversion. However, CE is recommended because it is shorter by one propagation delay than C̄S and is the faster input of the two.

Once STS goes high, signaling that a conversion has started, all convert-start commands have no effect until the conversion is complete. The output buffers cannot be enabled during a conversion.

1.3.5 Read Timing with Full Microprocessor Control

Figure 1-24 shows the read-cycle timing. During data reading, access time is measured from when CE and R/C̄ are both high. Access time is extended 10 ns if C̄S used to initiate a read.

1.3.6 Stand-Alone Operation

For systems that do not use or require full-bus interface (under microprocessor control, when a CE or C̄S signal or both are available), ICs can be operated in a stand-alone mode directly linked through dedicated ports. The R/C̄ input is used to control

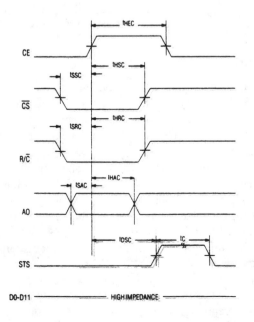

FIGURE 1-23
Convert-start timing for ADC IC (*Maxim New Releases Data Book,* 1992, p. 7-89)

the IC; the STS output provides an indication of conversion completion. To operate in stand-alone, $\overline{CS}$ and A0 hardwired low, with CE and 12/$\overline{8}$ wired high.

In stand-alone, R/$\overline{C}$ is set low to enable the three-state buffers. Conversion starts when R/$\overline{C}$ is set high. This allows either a high-pulse or low-pulse control signal to be applied at R/$\overline{C}$.

Figure 1-25 shows the timing for low-pulse control during stand-alone. In this mode, the outputs (in response to the falling edge of R/$\overline{C}$) are forced into the high-impedance state. The outputs return to valid-logic levels after the conversion is complete. The STS output goes high (following the R/$\overline{C}$ falling edge) and returns low when the conversion is complete.

Figure 1-26 shows the timing for high-pulse control during stand-alone mode. In this mode, the output data lines are enabled when R/$\overline{C}$ is high. The next conversion starts with the falling edge of R/$\overline{C}$. The data lines then return and remain in the high-impedance state until another high pulse is applied to R/$\overline{C}$.

1.3.7 Power Supply Requirements and Physical Layout

Figure 1-27 shows the recommended power-supply grounding. The ground-reference point for the on-chip reference is AGND (pin 9), which should be connected to the analog-reference point of the system. The analog and digital grounds should be connected together at the IC package to obtain maximum accuracy in high digital-noise environments. If the situation allows (low digital noise), the grounds can be connected to the most accessible ground-reference point, but the analog-power return is preferred.

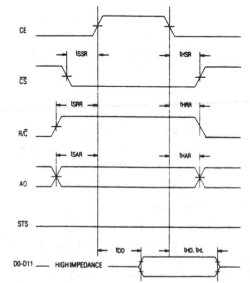

FIGURE 1-24
Read-cycle timing for ADC IC (*Maxim New Releases Data Book,* 1992, p. 7-89)

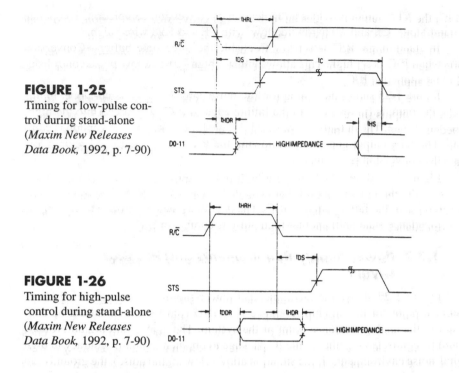

FIGURE 1-25

Timing for low-pulse control during stand-alone (*Maxim New Releases Data Book,* 1992, p. 7-90)

FIGURE 1-26

Timing for high-pulse control during stand-alone (*Maxim New Releases Data Book,* 1992, p. 7-90)

Figure 1-28 shows the recommended power-supply bypassing. The power supplies must be filtered, well regulated, and free from high-frequency noise. If they are not, unstable output codes can result, especially if switching spikes are present. Do not use switching power supplies for applications requiring 12-bit resolution. (A few mV of noise converts to several error counts in a 12-bit ADC.)

All power-supply pins should use supply decoupling capacitors connected with short leads to the pins. The VCC and VEE pins should be decoupled directly to AGND. A 4.7-μF tantalum capacitor in parallel with a 0.1-μF disk ceramic is recommended for decoupling.

ICs are designed for use on PC (printed circuit) boards. Wire-wrap boards are not recommended. Board layout should be made so that digital and analog signal lines are kept separated from each other as much as possible. Care should be taken not to run analog and digital lines parallel to each other or to run digital lines directly underneath the ICs (as is the case with any good analog-digital board layout).

1.3.8 Internal Reference

As shown in Fig. 1-18, the ICs have an internal buried-zener reference that provides a 10-V, low-noise, and low-temperature-drift output (at pin 8). An external reference voltage can also be used (at pin 10). When using ±15-V supplies, the internal reference can source up to 2 mA (in addition to the BiP$_{OFF}$ and REF$_{IN}$ inputs) over the

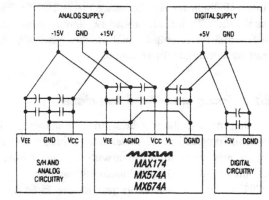

FIGURE 1-27
Recommended power-supply grounding for typical ADC IC (*Maxim New Releases Data Book,* 1992, p. 7-90)

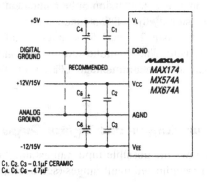

FIGURE 1-28
Recommended power-supply bypassing for typical ADC IC (*Maxim New Releases Data Book,* 1992, p. 7-91)

entire operating temperature range. With ± 12-V supplies, the reference can drive the BiP$_{OFF}$ and REF$_{IN}$ inputs over temperature, but *cannot drive an additional load.*

1.3.9 Analog Inputs

The input leads to AGND and 10V$_{IN}$ or 20V$_{IN}$ should be as short as possible to minimize noise pickup. Use shielded cables if long leads are required.

When the 20V$_{IN}$ is used as the analog input, load capacitance on the 10V$_{IN}$ pin must be minimized. Especially on the faster MAC174, leave the 10V$_{IN}$ pin open to minimize capacitance and to prevent linearity errors caused by inadequate settling time.

The amplifier used to drive the analog input must have low DC output impedance for low full-scale error. In addition, low AC output impedance is required because the analog input current is modulated at the clock rate during the conversion. (The output impedance of an amplifier is the open-loop output impedance divided by the log gain at the frequency of interest.)

The internal clock rate for the MAX714 is 2 MHz. This requires faster amplifiers such as the OP-27, AD711, or OP-42. The approximate internal clock rate for the MX574A is 600 kHz, with 1 MHz for the MX674A. An amplifier such as the MAX400 can accommodate these speeds.

1.3.10 Track-and-Hold Interface

The analog input to these ADCs must be stable to within ½ LSB during the entire conversion for specified 12-bit accuracy. This limits the input-signal bandwidth to a few Hz for sine-wave input (even with the faster MAX174). A track-and-hold amplifier should be used for higher bandwidth signals.

The STS output can be used to provide the *Hold* signal to the track-and-hold (T/H) amplifier. However, because the internal DAC is switched at about the same time as the conversion is initiated, the switching transients at the output of the T/H DAC switches might result in code-dependent errors. It is recommended that the *Hold* signal to the T/H amplifier precede a conversion or be coincident with the conversion start.

The first bit decision by the ADC is made approximately 1.5 clock cycles after the start of conversion. This is 2.5 µs, 1.5 µs, and 0.8 µs for the MX574A, MX674A and MAX174, respectively. The T/H hold settling time *must be less* than these times. Figures 1-29 and 1-30 show recommended T/H circuits for the MX574A/674A and MAX174, respectively.

1.3.11 Input Ranges and Digital Output Codes

Figure 1-31 shows the possible input ranges and "ideal" transitions voltages. End-point errors can be adjusted in all ranges (see Sections 1.3.12 and 1.3.13).

1.3.12 Typical Unipolar Input

Figures 1-32 and 1-33 show the ideal transfer function and input connections, respectively, for unipolar input operation. In many cases, the gain (full scale) and offset need not be calibrated. This is because all internal resistors of the ADCs are trimmed for absolute calibration. (The absolute accuracy for each grade of ADC is given in the data-sheet specifications.)

For a 0-V to +10-V input range, the analog input is connected between AGND and $10V_{IN}$, as shown in Fig. 1-33. For a 0-V to +20-V input range, connect the analog input between AGND and $20V_{IN}$. Note that these ADCs can easily handle input signals beyond the supply voltage.

Should a 10.24-V input range be required, connect a 200-ohm trimmer in series with $10V_{IN}$. For a full-scale input range of 20.48 V, use a 500-ohm trimmer in series with $20V_{IN}$. The nominal input impedance into $10V_{IN}$ is 5 k, and into $20V_{IN}$ is 10 k.

If the full-scale input and offset need not be trimmed, delete R1, R2, and the related wiring shown in Fig. 1-33. Connect BiP_{OFF} directly to AGND. Connect a 50-ohm metal-film resistor (±1%) between REF_{OUT} and REF_{IN}.

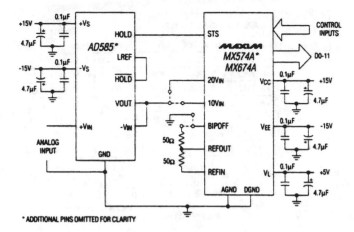

FIGURE 1-29 Recommended track-and-hold circuits for MX574A/MX674A (*Maxim New Releases Data Book,* 1992, p. 7-91)

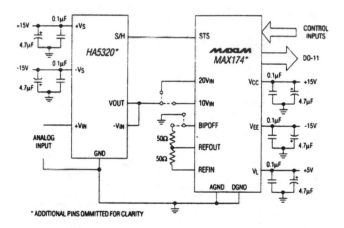

FIGURE 1-30 Recommended track-and-hold circuits for MAX174 (*Maxim New Releases Data Book,* 1992, p. 7-92)

If the full-scale input and offset must be trimmed, use the connections of Fig. 1-33 and proceed as follows. Adjust the offset first. Apply ½ LSB (see Fig. 1-31 for voltages) at the analog input and adjust R1 until the digital output code flickers between 0000 0000 0000 and 0000 0000 0001. Then apply full-scale less ³⁄₂ LSB (Fig. 1-31) at the analog input and adjust R2 until the output code changes between 1111 1111 1110 and 1111 1111 1111 (see Section 1.4 for a discussion of ADC trimming and adjustment).

ANALOG INPUT VOLTAGE (Volts)				DIGITAL OUTPUT	
0 to +10V	0 to +20V	±5V	±10V	MSB	LSB
+10.0000	+20.0000	+5.0000	+10.0000	1111 1111 1111	
+9.9963	+19.9927	+4.9963	+9.9927	1111 1111 1110*	
+5.0012	+10.0024	+0.0012	+0.0024	1000 0000 0000*	
+4.9988	+9.9976	-0.0012	-0.0024	0111 1111 1111*	
+4.9963	+9.9927	-0.0037	-0.0073	0111 1111 1110*	
+0.0012	+0.0024	-4.9988	-9.9976	0000 0000 0000*	
0.0000	0.0000	-5.0000	-10.0000	0000 0000 0000	

FIGURE 1-31

Input ranges and ideal digital output codes (*Maxim New Releases Data Book,* 1992, p. 7-9

Note 1: For unipolar input ranges, output coding is straight binary.
Note 2: For bipolar input ranges, output coding is offset binary.
Note 3: For 0V to +10V or ±5V ranges, 1LSB = 2.44mV.
Note 4: For 0V to +20V or ±10V ranges, 1LSB = 4.88mV.
* The digital outputs will be flickering between the indicated code and the indicated code plus one.

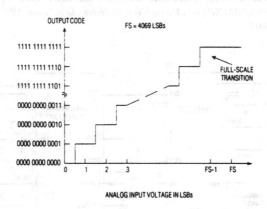

FIGURE 1-32

Ideal unipolar transfer function (*Maxim New Releases Data Book,* 1992, p. 7-93)

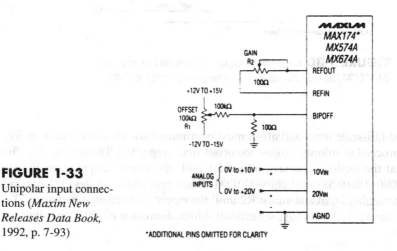

FIGURE 1-33

Unipolar input connections (*Maxim New Releases Data Book,* 1992, p. 7-93)

1.3.13 Typical Bipolar Input

Figures 1-34 and 1-35 show the ideal transfer function and input connections, respectively, for bipolar input operation. The full scale and offset need not be calibrated for all applications because of the internal-resistor accuracy and trimming. One or both of the trimmers (R1, R2) can be replaced with a 50-ohm ($\pm 1\%$) resistor if external trimming is not required. The analog input ranges can be either ± 5 V or ± 10 V, as needed.

If the full-scale and offset must be trimmed, use the connections of Fig. 1-35 and proceed as follows. Adjust the offset first. Apply ½ LSB above negative full-scale (see Fig. 1-31) at the analog input and adjust R1 until the digital output code flickers between 0000 0000 0000 and 0000 0000 0001. Then apply a voltage ¾ LSB below positive full-scale (Fig. 1-31) at the analog input and adjust R2 until the output code changes between 1111 1111 1110 and 1111 1111 1111.

1.4 Basic ADC/DAC Testing and Troubleshooting

This section is devoted to digital testing and troubleshooting basics. The testing and troubleshooting for specific data converter ICs are given in the related chapters. It is assumed that you are already familiar with digital troubleshooting at a level found in Lenk's *Digital Handbook* (McGraw-Hill, 1993). However, the following paragraphs summarize both testing and troubleshooting as they relate to ADC and DAC ICs.

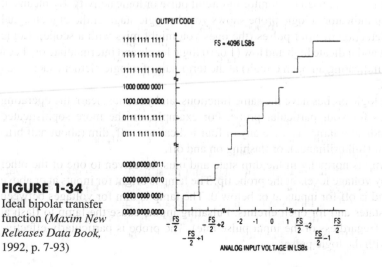

FIGURE 1-34
Ideal bipolar transfer function (*Maxim New Releases Data Book,* 1992, p. 7-93)

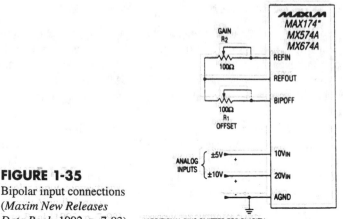

FIGURE 1-35

Bipolar input connections
(*Maxim New Releases
Data Book,* 1992, p. 7-93)

1.4.1 Digital Circuit Testing and Troubleshooting

Both testing and troubleshooting for the data converters in this book can be performed with conventional test equipment such as meters, generators, and scopes.
However, a logic or digital probe and a digital pulser can make life much easier if you
must regularly test and troubleshoot data converters (or any other digital device). So
we start with brief descriptions of the probe and pulser.

1.4.2 Logic or Digital Probe

Logic probes are used to monitor in-circuit pulse or logic activity. By means of
a simple lamp indicator, a logic probe shows you the logic stage of the digital signal
and allows detection of brief pulses (the ones you might miss with a scope). Logic
probes detect and indicate high and low (1 or 0) logic levels and intermediate or "bad"
logic levels (indicating an open circuit) at the terminals of a logic element such as an
ADC or DAC.

Not all logic probes have the same functions, and you must learn the operating
characteristics for your particular probe. For example, on the more sophisticated
probes, the indicator lamp can give any of four indications: off, dim (about half brilliance), bright (full brilliance), or flashing on and off.

The lamp is normally in the dim state and must be driven to one of the other
three states by voltage levels at the probe tip. The lamp is bright for inputs at or above
the 1 state and is off for inputs at or below 0. The lamp is dim for voltages between
the 1 and 0 states and for open circuits. Pulsating inputs cause the lamp to flash at
about 10 Hz (regardless of the input pulse rate). The probe is particularly effective
when used with the logic pulser.

1.4.3 Logic Pulser

The hand-held logic pulser (similar in appearance to the logic probe) is an in-circuit stimulus device that automatically outputs pulses of the required logic polarity, amplitude, current, and width to drive lines and other test points high or low. A typical pulser has several pulse burst and stream modes.

Logic pulses are compatible with most digital devices. Pulse amplitude depends on the equipment supply voltage, which is also the supply voltage for the pulser. Pulse current and pulse width depend on the load being pulsed. A switch controls the frequency and number of pulsers that are generated. A flashing LED (light-emitting diode) indicator on the pulser tip indicates the output mode.

The logic pulser forces overriding pulses into lines or test points, and it can be programmed to output single pulses, steady pulse streams, or bursts. The pulser can be used for ICs to be enabled or clocked. The circuit inputs also can be pulsed while the effects on the circuit outputs are observed with a logic probe.

1.4.4 General Digital Troubleshooting Sequence

The following troubleshooting tips are not limited to ADCs and DACs. They also apply to all digital circuits in which most of the components are contained within ICs.

1.4.5 Power and Ground Connections

The first step in tracing problems in a digital circuit with ICs is to check all power and ground connections to the ICs. Many ICs have more than one power and one ground connection. For example, the LTC1090 in Fig. 1-36 requires +5 V at the V_{CC} pin and ground at the V− pin. The IC also has both a digital ground (DGND) and an analog ground (AGND), as well as a common (COM) pin, and a minus-reference (REF−) pin that must be grounded. The REF+ pin must also be connected to 5-V power. If any of these power or ground connections is absent or abnormal, the IC cannot operate properly.

1.4.6 Reset, Chip-Select, and Start Signals

With all power and ground connections confirmed, check that all the ICs receive reset, chip-select, and start signals, as required. For example, the DAC-4881 in Fig. 1-37 requires a chip-select at pin 1 and address-decode signals at pins 2 and 28. Likewise, the ADC0808/0809 in Fig. 1-38 requires start, address latch enable (ALE), end-of-conversion (EOC), and output enable signals (Fig. 1-38c) from a microprocessor or control logic. If any of these signals is absent or abnormal (for example, incorrect amplitude, improper timing) circuit operation comes to an immediate halt.

In some cases, control signals to digital ICs are pulses (usually timed in a certain sequence), whereas other control signals are steady (high or low). If any of the

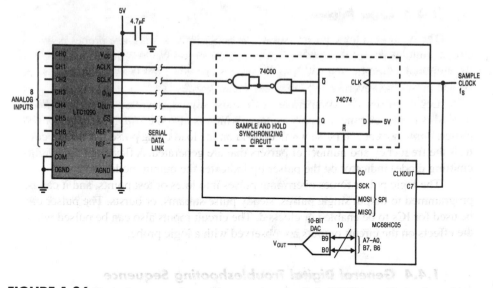

FIGURE 1-36 Typical power and ground connections for digital IC (Linear Technology, *Linear Applications Handbook,* 1993, p. DN2-1)

lines carrying the signals to the IC are open, shorted to ground, or to power (typically +5 V or +12 V), the IC will not function. So if you find an IC pin that is always high, always low, or apparently is connected to nothing (floating), check the PC traces or other wiring to that pin carefully. This applies to all control pins, unless the circuit calls for the control function to be steady (such as a steady +5 V on a pin to turn a circuit on). For example, if the DAC-4881 is connected as an 8-bit with complementary input DAC as shown in Fig. 1-37, the chip-select (pin 1) must receive a write (WR) signal, and the address-decode pins must receive address bits from the microprocessor.

1.4.7 Clock Signals

Most digital ICs require clocks. For example, Fig. 1-38c shows the clock periods for the ADC in Fig. 1-38. In this case, the clock comes from an external source. In other cases, the clock is part of the circuit. In general, the presence of pulse activity on any pin of a digital IC indicates the presence of a clock, but do not count on it. Check directly at the clock pins (all ICs that require a clock typically are connected to the same clock source).

It is possible to measure the presence of a clock signal with a scope or logic probe. However, a frequency counter provides the most accurate measurement. If any ICs do not receive required clock signals, the IC cannot function. On the other hand, if the clock is off frequency, all of the ICs might appear to have a clock signal, but the IC function can be impaired. Notice that crystal-controlled clocks do not usually drift far off frequency but can go into some overtone frequency (typically a third overtone) beyond the capacity of the IC.

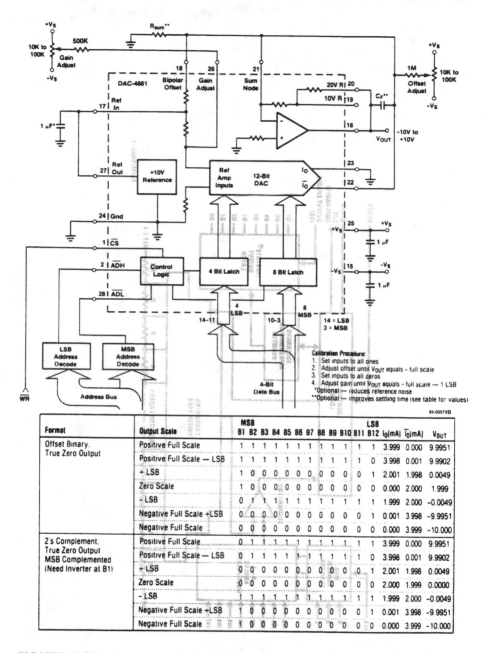

FIGURE 1-37 Typical control signals for digital IC (*Raytheon Semiconductor Data Book,* 1994, p. 6-48)

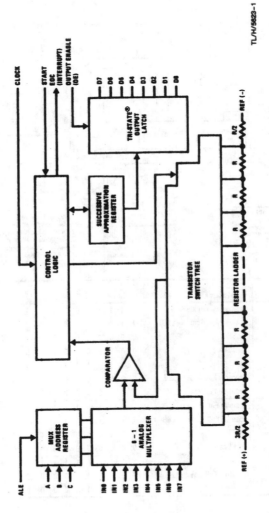

TL/H/5623–1

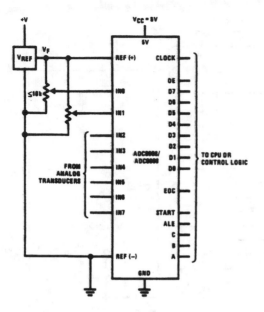

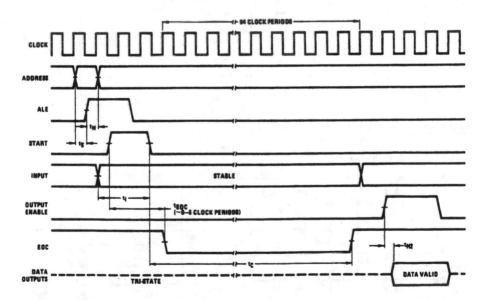

FIGURE 1-38 Typical control signals and timing diagram for digital IC (National Semiconductor, *Linear Applications Handbook*, 1994, p. 531/532)

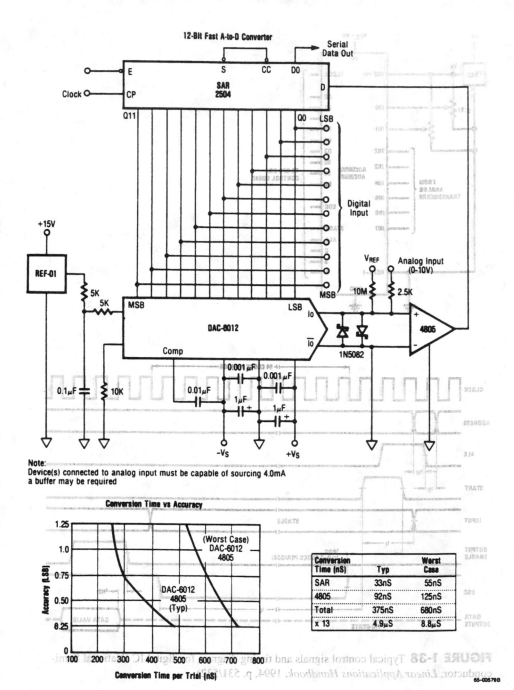

FIGURE 1-39 Typical test points for ADC circuit (*Raytheon Semiconductor Data Book*, 1994, p. 6-82)

1.4.8 Input-Output Signals

When you are certain that all ICs are good and have proper power and ground connections and that all control signals (such as reset, chip-select) and clock signals are available, the next step is to monitor all input and output signals at each IC. This can be done with either a scope or a probe. The following are some examples that apply to ADCs and DACs.

1.4.9 Basic ADC Testing and Troubleshooting

ADCs can be tested by applying precision voltages at the input and monitoring the output for corresponding digital values. For example, a fixed voltage between 0 and $+10$ V can be applied to the noninverting input of the 4805 in Fig. 1-39, and the corresponding value can be read out at the lines between the SAR-2504 and DAC-6012. The lines should go to $+5$ V for a digital 1 and to ground or 0 V for a digital 0. There is also a serial digital output at the D0 pin of the 2504. This output must be monitored with a scope.

The rate at which the conversions are made is controlled by the clock at the CP pin of the 2504. Notice that the start (S) pin of the 2504 is connected to the conversion-complete (CC) pin to provide continuous digital outputs for the analog input. In the circuit of Fig. 1-39, the start pin must receive a conversion command from an external source (typically a microprocessor), whereas the conversion-complete becomes an output to the microprocessor (indicating status and conversion complete or not complete).

If the output readings of the ADCs in this book are slightly off, try correcting the problem by means of adjustment. In the circuit of Fig. 1-39, the V_{REF} voltage (at the noninverting input of the 4805) can be varied for 0 (0 V at the analog input should make all digital outputs ground or 0 V). Also, REF-01 can be trimmed for full-scale output. This is done by connecting a 10-k pot between the output (pin 6) of the REF-01 and ground (pin 4). The wiper of the pot is connected to trim (pin 5) of the REF-01. The accuracy of this circuit depends on the precision of the two 5-k resistors between the REF-01 and pin 14 of the DAC-6012 and on the 2.5-k resistor at the analog input.

1.4.10 Basic DAC Testing and Troubleshooting

DACs can be tested by applying digital inputs and monitoring the output for corresponding voltages. For example, the B1 through B10 inputs of the DAC-10 can be connected to ground (for a 0) or to $+5$ V (for a 1), and the output can be monitored with a precision voltmeter at pins 2 and 4 in the circuit of Fig. 1-40. If both voltages are slightly off, suspect the 2.00-mA references. If one of the output voltages is slightly off, suspect the corresponding 1.25-k resistors. If the output voltages are absent or way off, suspect the DAC-10

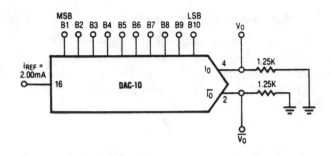

FIGURE 1-40

Typical test points for DAC circuit (*Raytheon Semiconductor Data Book,* 1994, p. 6-34)

	B1 B2 B3 B4 B5 B6 B7 B8 B9 B10	I$_0$mA	I̅$_0$mA	V$_0$	V̅$_0$
Full Scale	1 1 1 1 1 1 1 1 1 1	3.996	0.000	-4.995	-0.000
Half Scale +LSB	1 0 0 0 0 0 0 0 0 1	2.004	1.992	-2.505	-2.490
Half Scale	1 0 0 0 0 0 0 0 0 0	2.000	1.996	-2.500	-2.495
Half Scale −LSB	0 1 1 1 1 1 1 1 1 1	1.996	2.000	-2.495	-2.500
Zero Scale +LSB	0 0 0 0 0 0 0 0 0 1	0.004	3.992	-0.005	-4.990
Zero Scale	0 0 0 0 0 0 0 0 0 0	0.000	3.996	-0.000	-4.995

CHAPTER 2

Data-Converter Terms and Design Characteristics

Much of the basic design information for a particular IC data converter can be obtained from the data sheet. Likewise, a typical data sheet describes a few specific applications for the data converter. However, converter data sheets often have two weak points. First, they assume that everyone understands all the terms used. Of more importance, the data sheets do not show *how the listed parameters relate to design problems.* To further complicate the situation, each manufacturer has a separate system of data sheets. It is impractical to discuss all data sheets herein. Instead, we discuss typical information found on the converter data sheets, and see how this information affects simplified design. We start with definitions of some basic terms.

2.1 Resolution and Accuracy

The terms *resolution* and *accuracy* are often interchanged (although incorrectly). In a DAC, *resolution* describes the smallest standard incremental change in output voltage. In an ADC, *resolution* is the amount of input voltage change required to increment the output between one code change and the next adjacent code change. Both definitions of resolution differ from the definition of *accuracy*, which is the absolute error incurred in measurement of a signal. In many data converters, accuracy does not match resolution. Let us consider some examples.

A converter with n switches can resolve one part in 2^n. The least-significant increment is then 2^{-n}, or one LSB. In contrast, the MSB carries a weight of 2^{-1}. Resolution can be expressed in percentage of full scale, or in binary bits. For example, an ADC with 12-bit resolution can resolve one part in 2^{12} (one part in 4096) or 0.0244% of full scale. A converter with 10 V full scale can resolve a 2.44-mV input change. Likewise, a 12-bit DAC shows an output-voltage change of 0.0244% of full scale when the binary input code is incremented one binary bit (1 LSB). Resolution is a design parameter rather than a performance specification (because resolution says nothing about accuracy or linearity).

A *linearity* specification (see Section 2.2) is sometimes used in place of accuracy in data converters, because linearity is more descriptive. When used, an accuracy specification describes the worst-case deviation of a DAC output voltage from

39

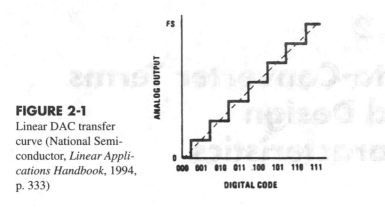

FIGURE 2-1

Linear DAC transfer curve (National Semiconductor, *Linear Applications Handbook*, 1994, p. 333)

a straight line drawn between zero and full scale. For example, a 12-bit DAC cannot have a conversion accuracy better than ±½ LSB or ±1 part in 2^{12+1} (±0.0122% of full scale) because of the finite resolution. This is the case in Fig. 2-1, if there are no other errors. Note that ±0.0122% full scale represents a deviation from 100% accuracy. Therefore, accuracy should be specified as 99.9878%. However, most data sheets would use 0.0122% as an accuracy specification, rather than an inaccuracy (tolerance or error) specification (to further confuse users).

In an ADC, *accuracy* describes the difference between the actual input voltage and the full-scale weighted equivalent of the binary output code, including all other errors (see Section 2.3). For example, if a 12-bit ADC is said to be ±1 LSB accurate, this is equivalent to ±0.0245%, or twice the minimum possible quantizing error of 0.0122%. In effect, an accuracy specification describes the maximum sum of all errors.

2.2 Linearity

Linearity specifications describe the departure from a linear transfer curve for an ADC or DAC. Linearity error does not include quantizing, offset, zero, or scale errors (see Section 2.3). Thus a specification of ±½ LSB linearity implies error, in addition to the inherent ±½ LSB quantizing or resolution error. This is shown in Fig. 2-2, in which a linearity error allows one or more of the steps to be greater or less than the ideal shown.

One of the problems with linearity specifications (or nonlinearity specifications, if you prefer) is that there two testing approaches. Figure 2-3 shows how these two approaches (the *best-straight-line* and *endpoint-fit*) produce different specifications for an ADC.

The best-straight-line approach makes no claim about zero error, full-scale error, or transfer-function slope but simply quantifies (in LSBs or percentages) deviation from the straight line that best approximates the transfer function. In effect, the best-straight-line method provides the lowest (or "best looking") number. No points on the line are defined before the test. The result is a pure linearity specification that includes no other errors.

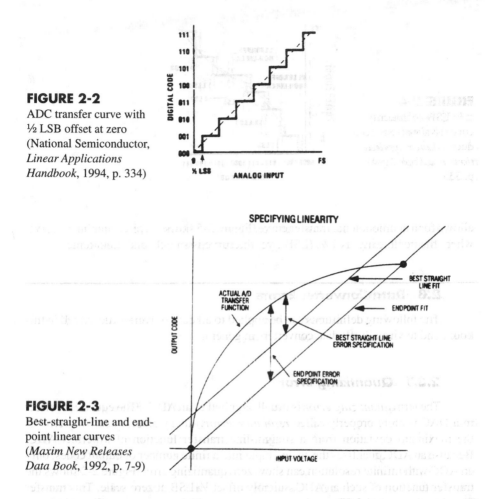

FIGURE 2-2
ADC transfer curve with
½ LSB offset at zero
(National Semiconductor,
*Linear Applications
Handbook*, 1994, p. 334)

FIGURE 2-3
Best-straight-line and end-
point linear curves
(*Maxim New Releases
Data Book*, 1992, p. 7-9)

   The endpoint-fit approach presets the ideal line between the measured end-
points of the data-converter transfer function. Deviations are measured without
adjusting the position of the line for any optimum fit. As a result, the endpoint-fit
linearity number is usually larger than that of the best-straight-line approach. However, both methods are valid ways of representing linearity (also called *integral nonlinearity* [INL] on some data sheets).
   Figure 2-4 shows a 3-bit DAC transfer curve with no more than ±½ LSB nonlinearity, yet one step is of zero amplitude. This is still within the specification,
because the maximum deviation from the ideal straight line is ±1 LSB (½ LSB resolution error, plus ½ LSB nonlinearity).
   With any linearity error, there is *differential nonlinearity* (see Section 2.4). A
±½ LSB linearity specification guarantees monotonicity (see Section 2.5) and an
equal or better than ±1 LSB differential nonlinearity (DNL). In the example of
Fig. 2-4, the code transition from 100 to 101 is the worst possible nonlinearity (1 LSB
high at code 100 to 1 LSB low at 110). Any fractional nonlinearity beyond ±½ LSB

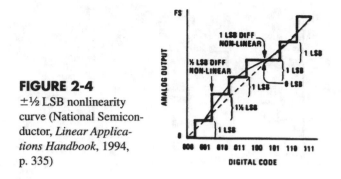

FIGURE 2-4
±½ LSB nonlinearity
curve (National Semicon-
ductor, *Linear Applica-
tions Handbook*, 1994,
p. 335)

allows for a nonmonotonic transfer curve. Figure 2-5 shows a typical nonlinear curve, where the nonlinearity is 1 ¼ (LSB), yet the curve is smooth and monotonic.

2.3 Data Converter Errors

The following definitions can be applied to all data converters described in this book, and to virtually all data converters in general.

2.3.1 Quantizing Error

The term *quantizing error* is usually applied to an ADC. (The equivalent effect in a DAC is more properly called *resolution error*.) In any case, quantizing error is the maximum deviation from a straight-line transfer function of a perfect ADC. Because an ADC quantizes the analog input into a finite number of output codes, only an ADC with infinite resolution can show zero quantizing error. Figure 2-2 shows the transfer function of such an ADC, suitably offset ½ LSB at zero scale. This transfer function shows ±½ LSB maximum output error. If there were no offset, the error would be $^{-1}_{+0}$ LSB as shown in Fig. 2-6. For example, a perfect 12-bit ADC shows

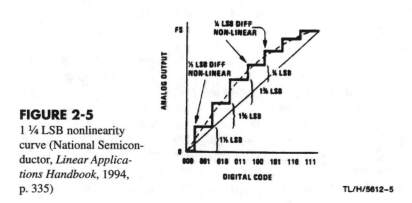

FIGURE 2-5
1 ¼ LSB nonlinearity
curve (National Semicon-
ductor, *Linear Applica-
tions Handbook*, 1994,
p. 335)

TL/H/5612-5

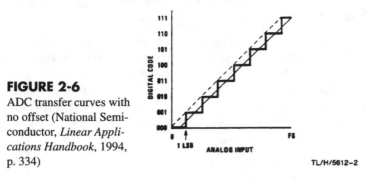

FIGURE 2-6
ADC transfer curves with no offset (National Semiconductor, *Linear Applications Handbook*, 1994, p. 334)

TL/H/5612-2

a $\pm\frac{1}{2}$ LSB error of $\pm0.0122\%$, whereas the quantizing error of an 8-bit ADC is $\pm\frac{1}{2}$ part in 2^8 or $\pm0.195\%$ of full scale.

2.3.2 Scale Error

Scale error (also known as *full-scale error*) is the departure from design output voltage of DAC for a given input code (usually full-scale code). Figure 2-7 shows a transfer function with a linear 1 LSB scale error. In an ADC, scale error is the departure of actual input voltage from design input voltage for a full-scale output code.

Scale errors can be caused by errors in reference voltage, ladder resistor values, or amplifier gain (see Chapter 1) and can be corrected by means of adjustments in output amplifier gain or reference voltage. For example, if the transfer curve resembles that of Fig. 2-5, a scale adjustment at ¾ scale could improve the overall ± accuracy compared with an adjustment at full scale.

2.3.3 Gain Error

In an ADC, gain error is essentially the same as scale error. In the case of a DAC with current and voltage-mode outputs, the current output can be to scale, but the voltage output might show some gain error. In any case, the amplifier-feedback resistors can be trimmed to correct gain-error problems.

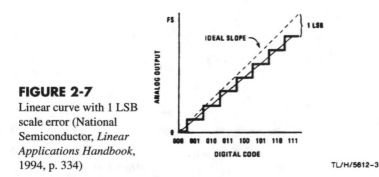

FIGURE 2-7
Linear curve with 1 LSB scale error (National Semiconductor, *Linear Applications Handbook*, 1994, p. 334)

TL/H/5612-3

2.3.4 Offset Error

Offset error (also known as *zero error*) is the output voltage of a DAC, with zero-code input. In an ADC, offset error is the required mean value of input voltage to set a zero-code output. Figure 2-8 shows a DAC transfer curve with ½ LSB offset at zero.

Offset error is usually caused by an input-offset voltage or current of the amplifier or comparator within the converter IC (see Chapter 1) and can be trimmed to zero with an external offset-zero potentiometer. Offset error can be expressed in percentage of full-scale or in a fraction of an LSB.

2.3.5 Hysteresis Error

The term *hysteresis error* usually applies to an ADC. The error causes the voltage at which an ADC code transition occurs to depend on the direction from which the transition is approached. Hysteresis error in an ADC is usually caused by hysteresis in the internal comparator (see Chapter 1). Excessive hysteresis is reduced by design of the converter. However, some slight hysteresis is inevitable. For our purposes, hysteresis is objectionable if it approaches ½ LSB.

2.3.6 Trimming Data-Converter Errors

As discussed in Chapter 1, some data-converter ICs provide pins for external trimming to offset any errors or to achieve a desired accuracy. In other cases, the accuracy is built in (usually at a higher cost). There are also instances in which greater accuracy or offset is simply not needed. Of course, if space is the ultimate consideration, you must use the converter without external components. Unfortunately, this might mean using a more expensive converter to get the required accuracy. The following points should be considered when pondering the trimmed as opposed to untrimmed accuracy tradeoff.

A typical low-cost 12-bit ADC (such as a MAX172) resolves a 5-V full-scale input range to 5V/4096, or 1.2 mV (1 LSB), but is not accurate to this level. The untrimmed full-scale error limit of the B-grade IC is 15 LSB, with an offset limit of

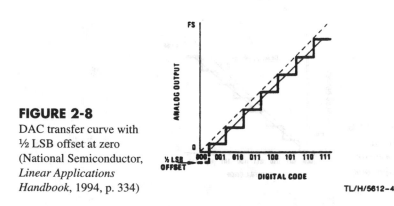

FIGURE 2-8
DAC transfer curve with ½ LSB offset at zero (National Semiconductor, *Linear Applications Handbook*, 1994, p. 334)

TL/H/5612–4

6 LSB. These guarantees allow reduced cost *without sacrificing linearity,* which is guaranteed at ½ LSB for the A-grade or 1 LSB for the B-grade device. Twelve-bit linearity is maintained (even on the lowest grade).

In simplified design, unadjusted full-scale and offset errors are often not critical. This is because such errors are constant, and errors in other parts of the signal path are often trimmed. As a general guideline, ADC accuracy and offset specifications do not need full 12-bit precision when:

1. Only *signal changes,* not absolute voltage levels, are of interest.
2. One is measuring transducers or sensors that do not have precise accuracy specifications (for example, the MAX172 B-grade 15-LSB full-scale error = 0.37%).
3. System calibration occurs elsewhere, either with manual trims or through a microprocessor or controller.
4. The ADC operates in a closed-loop control system in which gain and offset errors affect only loop dynamics and not accuracy.

From a simplified-design standpoint, if you use high-resolution ADCs with the required untrimmed accuracy, such accuracy must be 1 LSB or better. If not, you do not really eliminate trims but only reduce the required range of the trims.

2.4 Differential Nonlinearity

In an ADC, DNL indicates that the device is monotonic (see Section 2.5) or has no missing codes. DNL is the deviation of the analog span of each ADC output code from the ideal 1 LSB value. (Figure 2-9 shows some typical DNL errors.) For example, a DNL specification of ½ LSB means that a code is at least ½ LSB but no more than 1.5 LSB wide. If DNL is less than 1 LSB, no missing codes is assured. (Some data sheets list DNL and no missing codes as separate characteristics.)

In a DAC, DNL indicates the difference between actual analog voltage change and the ideal (1 LSB) voltage change at any code change. For example, a DAC with

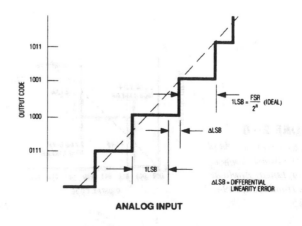

FIGURE 2-9
Typical DNL errors
(*Maxim New Releases
Data Book*, 1992, p. 7-10)

a 1.5 LSB step at a code change is said to show a DNL of ½ LSB (see Figs. 2-4 and 2-5). DNL can be expressed in LSB or percentage of full-scale.

In simplified design, DNL specifications are as important as linearity specifications because the *apparent quality* of a data-converter curve can be markedly affected by DNL, even though linearity is good. For example, Fig. 2-4 shows a curve with ±½ LSB linearity and ±1 LSB DNL. Figure 2-5 shows a curve with +1 ¼ LSB linearity and ±½ LSB DNL. In many applications, the curve of Fig. 2-5 would be preferred over that of Fig. 2-4 because the Fig. 2-5 curve is smoother. (In simple terms, DNL describes the smoothness of the curve and is thus of great importance to the user.)

Figure 2-10 shows an exaggerated example of DNL in which the DNL is ±2 LSB and the linearity specification is ±1 LSB. This results in a transfer curve with a grossly degraded resolution. For example, the normal 8-step curve is reduced to three steps, and a 16-step curve (4-bit converter) with only 2 LSB DNL is reduced to six steps (a nonexistent 2.6-bit device!).

Unfortunately, DNL is not always listed on data sheets by all manufacturers. One reason for this is that DNL is difficult to measure on a production-line basis. Listed or not, DNL can be as much as twice nonlinearity, but never more.

2.5 Monotonicity

Figure 2-11 shows a nonmonotonic DAC transfer curve. For the curve to be nonmonotonic, the linearity error must exceed ±½ LSB by no matter how little. The greater the linearity error, the more significant the negative step might be. On the other hand, a monotonic curve has no change in sign of the slope. All incremental elements of a monotonically increasing curve have positive or zero (but never negative) slope. The converse is true for decreasing curves.

A converter showing more than ±½ LSB nonlinearity can still be monotonic up to a certain point, but not beyond that point. For example, a 12-bit DAC with ±½-bit linearity to 10 bits (not a true ±½ LSB) will be monotonic at 10 bits, but might or might not be monotonic at 12 bits (unless tested and guaranteed to be 12-bit monotonic). Finally, a nonmonotonic converter might be acceptable for some applications.

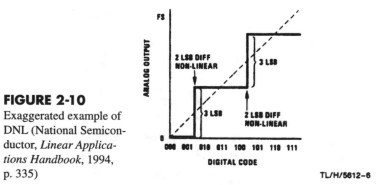

FIGURE 2-10

Exaggerated example of DNL (National Semiconductor, *Linear Applications Handbook*, 1994, p. 335)

TL/H/5612-6

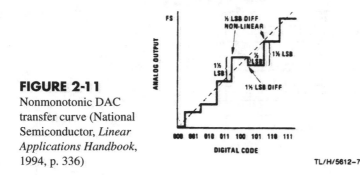

FIGURE 2-11

Nonmonotonic DAC
transfer curve (National
Semiconductor, *Linear
Applications Handbook*,
1994, p. 336)

TL/H/5612-7

However, nonmonotonic converters *are disastrous in closed-loop servo systems,*
including DAC-controlled ADCs.

2.6 Settling Time and Slew Rate

As discussed in Section 1.2.7, settling time is a critical factor in converter performance and is often listed along with slew rate. Settling time is the elapsed time
after a code transition for DAC output to reach final value within specified limits,
usually $\pm\frac{1}{2}$ LSB. Slew rate is an inherent limitation of the output amplifier in a DAC,
and functions to limit the output voltage rate-of-change after code transitions.

Settling time is often summed with slew rate to obtain total elapsed time for the
output to settle to final value. This is shown in Fig. 2-12, which delineates the part of
total elapsed time considered to be slew rate and the part that is settling time. As
shown, the total time is greater for a major code change than a minor code change
because of amplifier slew limitations. However, settling time also can be different,
depending on amplifier overload-recovery characteristics.

2.7 Conversion Rate

Both settling time and slew rate (as well as delay in counting circuits, ladder
switches, and comparators) affect conversion rate (the speed at which an ADC or
DAC can make repetitive data conversions). On some data sheets, conversion time is
specified as a number of conversions per second. Other data sheets list conversion
rate as the number of microseconds required to complete one conversion (including
the effects of settling time). Some data sheets specify conversion rate for something
less than full resolution, thus showing a misleading (high) rate.

2.8 Temperature Coefficient and Long-Term Drift

Temperature coefficient (TC) of the various components of a DAC or ADC can
produce (or increase) any of several errors when the operating temperature varies.
Zero-scale offset error can change because of the amplifier and comparator input-
offset TC. Scale error can occur because of shifts in the reference, changes in ladder

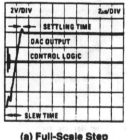

(a) Full-Scale Step

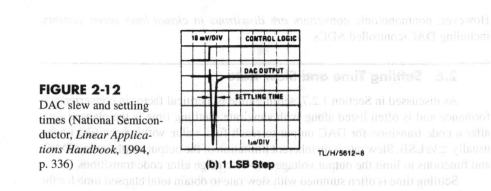

FIGURE 2-12
DAC slew and settling
times (National Semicon-
ductor, *Linear Applica-
tions Handbook*, 1994,
p. 336)

(b) 1 LSB Step

TL/H/5612-8

resistance, change in beta in current switches, or drift in amplifier gain-set resistors. Linearity and monotonicity of the DAC can be affected by different temperature drifts of the ladder resistors and switches. Many other characteristics can be affected by TC. In fact, with the possible exception of resolution and quantizing error, all data-converter characteristics can be affected by temperature changes.

Long-term drift, caused mainly by resistor and semiconductor aging, can affect all characteristics that temperature change can affect. The characteristics most commonly affected by long-term drift are linearity, monotonicity, scale, and offset. Scale changes because reference-voltage changes are usually the most important long-term change or drift problem.

2.9 Overshoot and Glitches

Both overshoot and glitches are essentially DAC problems. However, because most ADCs contain a DAC, ADCs also can be affected. Overshoot and glitches occur whenever a code transition occurs in a DAC. There are two causes. The current output of a DAC contains switching glitches because of possible asynchronous switching of the bit currents (expected to be worst at ½-scale transition when all bits are switched). Although such glitches are of extremely short duration, they could be ½ scale in amplitude. Although the glitches are generally attenuated at the DAC voltage output (because the amplifier is unable to slew at a very high rate), the glitches are coupled around the amplifier through the feedback network. In addition to the glitches, the output amplifier introduces some overshoot and some noncritical damped ringing. These problems can be minimized, but not entirely eliminated (except at the expense of slew rate and settling time, which are much more important).

2.10 Power Supply Rejection

Supply rejection relates to the ability of a DAC or ADC to maintain scale, offset, TC, slew rate, and linearity when the supply voltage is varied. The reference voltage must be constant (unless considering a multiplying DAC). Most affected are current sources (affecting linearity and scale) and amplifiers or comparators (affecting offset and slew rate). Supply rejection is usually specified as a percentage of full-scale change at or near full scale (at 25°C).

2.11 Input Impedance and Output Drive

Input impedance of an ADC describes the load placed on the analog source. Output drive describes the digital load-driving capability of an ADC or the analog load-driving capacity of a DAC. Output drive is usually given as a current level or a voltage output into a given load.

2.12 Clock Rate

For an ADC, clock rate is the minimum or maximum pulse rate at which the counters can be driven. There is a fixed relation between minimum conversion rate and clock rate that depends on converter accuracy and type. All factors that affect the conversion rate of an ADC limit the clock rate.

2.13 Data-Converter Codes

Data converters use most of the codes found in other digital equipment. Several types of DAC input, or ADC output, codes are in common (Fig. 2-13). Each code has advantages, depending on the system to be interfaced with the converter. The following is a summary of the most common data-converter codes.

2.13.1 Natural Binary

Natural (or simple) binary is the usual 2^n code with 2, 4, 8, 16, . . . 2^n progression. An input or output high, or 1, is considered a signal, whereas 0 is considered an absence of signal. This is a positive-true binary signal. Zero scale is all zeros, and full scale is all ones.

2.13.2 Complementary Binary

Complementary (or inverted) binary is the negative-true binary system. It is identical to natural binary except that all binary bits are inverted. Thus zero scale is all ones, and full scale is all zeros.

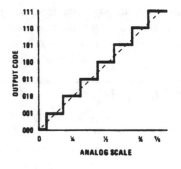

(a) Zero to + Full-Scale

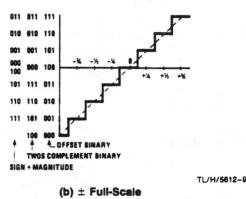

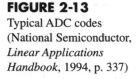

FIGURE 2-13
Typical ADC codes
(National Semiconductor,
*Linear Applications
Handbook*, 1994, p. 337)

TL/H/5612-9

(b) ± Full-Scale

2.13.3 Binary Coded Decimal

Binary coded decimal (BCD) is the representation of decimal numbers in binary form. It is useful in ADC systems intended to drive decimal displays. The advantage of BCD over decimal is that only four lines are needed to represent 10 digits. The disadvantage of BCD is that a full 4 bits can represent 16 digits, whereas only 10 digits are represented in BCD. The full-scale resolution of a BCD-coded system is less than that of a binary-coded system. For example, a 12-bit BCD system has a resolution of only one part in 1,000 compared with one part in 4,098 for a binary system. This represents a loss of resolution of more than 4:1.

2.13.4 Offset Binary

Offset binary is a natural binary code, except that it is offset (usually ½ scale) to represent negative and positive values. Maximum negative scale is represented as all zeros and with maximum positive scale as all ones. Zero scale (actually center scale) is then represented as a leading 1 and all remaining 0s.

2.13.5 Two's Complement Binary

Two's complement binary is widely used to represent negative values. With two's complement, zero and positive values are represented as in natural binary, and all negative values are represented in a two's complement form (see Fig. 2-13). That is, the two's complement of a number represents a negative value so that interface to a computer or microprocessor is simplified.

The two's complement is formed by complementing each bit and then adding a 1. Any overflow is neglected. For example, the decimal number -8 is represented in two's complement as follows: start with a binary code of decimal 8 (off scale for $\pm$ representation in 4 bits, so not a valid code in the $\pm$ scale of 4 bits), which is 1000. Complement this number to 0111 and add 0001 to get 1000.

The offset-binary representation of the $\pm$ scale differs from the two's complement representation only in that the MSB is complemented. The conversion from offset binary to two's complement requires that only the MSB be inverted.

2.13.6 Sign Plus Magnitude

As shown in Fig. 2-13, sign plus magnitude contains polarity information in the MSB. (When the MSB is 1, the number is negative.) All other bits represent magnitude only. One code is used up providing a double code for zero (000 or 100). Sign plus magnitude code is used in certain instrument applications (such as digital voltmeters [DVMs]) and in audio circuits. The advantage in both applications is that only one bit has to be changed for small-scale changes when the value is near zero and plus and minus scales are symmetric.

CHAPTER **3**

Practical Design Considerations

Now that we know how data converters work and we have settled on some basic terms, let us discuss practical design considerations. The remaining chapters in this book describe how a specific IC data converter can be used. The information in this chapter applies to all data converters, both ADC and DAC, and is included primarily so that the reader can select a converter to suit a specific system need. This should keep overspecification (with the usual high cost) to a minimum.

When reading this chapter, and all subsequent chapters, keep in mind that specific parameters, test conditions, test circuits, and even definitions might vary from manufacturer to manufacturer. For practical production reasons, parameters might not be tested in the same way for all converter types, even those supplied by the same manufacturer. However, using the information in this and remaining chapters, you should be able to sort out and understand specifications (from any manufacturer) that apply to your application.

3.1 Digital Control Signals

Data converters are digital devices and thus require digital control signals. Each ADC must accept or provide digital control signals telling it or the external system what to do and when to do it. Control signals should be compatible with one or more types of logic (TTL, CMOS, ECL) in common use. Control-signal timing must be such that the converter (or system) accepts the signals. The following is a summary of control signals found in most practical applications.

3.1.1 Start Conversion

Start conversion (SC) is a digital signal to an ADC that initiates a single conversion cycle. An SC signal typically must be present at the fall (or rise) of the clock waveform to initiate the cycle. A DAC needs no SC signal. However, an SC signal can be provided to gate digital inputs to a DAC.

3.1.2 End of Conversion

End of conversion (EOC) is a digital signal from an ADC that informs the external system that the digital output data bits are valid. Typically, an EOC output can be connected to an SC input to cause the ADC to operate in a continuous-conversion mode. In noncontinuous systems, the SC signal is a command from the system to the ADC. A DAC does not supply an EOC signal.

3.1.3 Clock Signals

DACs do not require clock signals. However, clocks are required by an ADC (or must be generated in the ADC IC) to control counting or successive-approximation registers. The clock controls the conversion rate or speed (within the limitations of the ADC IC).

3.2 Voltage References

As discussed in Chapter 1, data converters require a voltage reference. Most present-day IC converters have internal references. Figure 3-1 shows some typical examples. The suitability of a reference for a particular application depends mostly on *accuracy* and *drift*. As shown in Fig. 3-1, typical accuracy ranges from 0.2% to 1.63% for different ADCs.

From a simplified-design standpoint, reference accuracy and temperature drift relate directly to *full-scale errors*. If absolute accuracy is required, a precision reference must be used. If the requirements are too demanding for the internal reference, many converter ICs also accept external references (and there are converter ICs without internal references). Keep in mind that references with higher untrimmed accuracy cost more. So, if you are concerned primarily with stability and repeatability over a given temperature range, look for a reference with a low temperature coefficient (TC), but with less accuracy and a lower cost. Also remember that the reference on some IC converters can be trimmed to a given accuracy. However, this involves external components (with added cost and space consumption).

3.2.1 Ratiometric Data Conversion

One solution to the reference-accuracy problem is to use a ratiometric system in which the same reference drives both the converter and the external system or device. With ratiometric, the actual reference-voltage value (along with the error and drift) does not affect the result. If the reference voltage increases with temperature (or vice versa) both the converter and system change by the same ratio. As a result, the reference voltage need not show perfect stability over time and temperature. Either the data converter or an external source can provide the reference. In many applications, the power-supply voltage can be used as the reference (if the system supply is free from noise).

A/D PART NO.	RESOLUTION (BITS)	REF (%) TOLERANCE	REF VOLTAGE	REF DRIFT (ppm/°C)	
				Typ	Max
MAX150/4/8	8	1.2	2.5	40	70
MAX165/6	8	1.63	1.23	40	70
MAX151	10	0.75	4.00	—	60
MAX173	10	0.95	-5.25	40	—
MAX177	10	0.4	-5.00	—	45
MAX162	12	0.95	-5.25	20	—
MAX163/4/7	12	0.4	-5.00	—	25
MAX170	12	0.95	-5.25	20	—
MAX171	12	0.95	-5.25	—	40
MAX172	12	0.95	-5.25	40	—
MAX174	12	0.2	10.00	—	—
MAX178/182	12	0.3	5.00	10	40
MAX180/181	12	0.4	-5.00	—	25

FIGURE 3-1 Typical ADC internal reference accuracy and drift (*Maxim New Releases Data Book*, 1992, p. 7-10)

Of course there are applications in which absolute accuracy is required. For example, a direct voltage measurement using a digital meter (based on use of an ADC) is absolute and requires absolute reference accuracy. A ratiometric system should not be used in such cases.

A ratiometric system can be used when the information to be converted is the output signal from a bridge-type transducer fed into an ADC. The bridge output (for example, from a load cell or pressure transducer) is a function of the quantity (weight, pressure) and the bridge-excitation voltage. If the converter reference (or an external reference) is used for bridge excitation, variations in reference voltage will be applied equally (and in the same ratio).

3.3 Multichannel Conversion

When many channels of data must be converted, special problems arise in selecting a data-converter IC or system. The following is a summary of these problems and possible solutions.

3.3.1 Data Converters with Built-in Multiplexers

Unless the information to be converted is in a system with unlimited space (and high cost can be tolerated), some form of multiplexer must be used. For example, when several analog inputs are measured, a signal multiplexer is usually used at the ADC input. The multiplexer can be external, or an ADC with internal multiplexer (such as shown in Fig. 3-2) can be used. Such ADC ICs also offer some performance advantages over an external multiplexer ADC combination, besides providing the advantage of component savings. Multiplexer-error contributions and

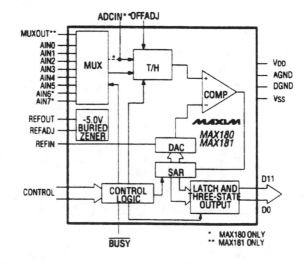

FIGURE 3-2

12-bit ADC with on-chip multiplexer, track-and-hold, and internal reference (*Maxim New Releases Data Book*, 1992, p. 7-11)

settling-time delays are eliminated from system specifications because they are included in the ADC error and timing limits. For example, the MAX180 shown in Fig. 3-2 has a 7.5-μs conversion time. This includes the time used by the multiplexer, the track-and-hold (T/H), and the ADC function.

Another advantage of the single-chip ADC with built-in multiplexer is that the interface is simplified. Only one device has to be addressed. In the IC of Fig. 3-2, the same operation can be used to select a channel and start a conversion. This cuts down on decoding logic and saves power.

3.3.2 Data Converters with External Multiplexers

Although multiple-input ADCs provide both convenience and performance advantages, such ICs have limitations that a separate ADC and multiplexer can sometimes overcome. An example is when the *relative timing of multiplexed signals is critical.* A typical ADC with multiplexer (one-chip or not) measures each channel in sequence—channel 1 is read long before channel 8, and so on. In some systems, this might add error if the signals of interest occur at the same time (the relative phase might contain information that the channel scanning distorts).

One solution to the timing problem is to *scan and convert at a fast rate* so that the time delay between channels is reduced to insignificant levels. This requires higher clock speed (along with the usual higher cost and possibly higher power consumption). Another solution is to *scan or sample all input channels at the same time,* then multiplex through T/H circuits before conversion. This eliminates scan-timing errors. (CMOS converters work well with such an approach.)

Any data-converter system with multiple inputs works best when *all input levels* have a similar dynamic range. If filtering or widely different channel gain is needed in a system, an IC with internal multiplexing usually requires that each channel has its own signal-conditioning circuit (a repetitive and expensive solution). This problem can be overcome with a *single filter and programmable amplifier* between an external multiplexer output and an ADC input.

The IC of Fig. 3-2 uses this technique, but does not require an external multiplexer. The output of the six-channel multiplexer (mux) and the input of the ADC are on separate pins, so a lowpass filter can be connected between the two. (A lowpass filter, such as the MAX270, provides selectable fourth-order corner frequencies between 1 kHz and 20 kHz for different channels.)

The final advantage of a separate multiplexer is that *greater fault protection* can be provided. Most present-day ADC ICs do not have high-voltage fault protection because structures are too large. Even when input-limiting resistors and external clamp diodes are used on each channel, there can be problems. (Crosstalk, caused when inactive channels are driven beyond the supply voltage, is a typical problem.) If such small, but significant, errors cannot be tolerated, an external multiplexer with high fault protection is the best solution. For example, the MAX358 or MAX378 fault-protected analog multiplexers can withstand up to 70-V overloads without spilling overdrive signals through to selected channels (crosstalk).

3.4 Track and Hold

Many ADCs include T/H functions as part of the internal circuit (such as the IC of Fig. 3-2). The T/H prevents the input from moving during the successive-approximation conversion process. As a (theoretic) guideline, a T/H function is required if the input signal changes during the time required for conversion to ½ LSB.

It is often assumed that T/H is not required for high-speed ADCs because conversion is complete before any significant changes occur. Although this is generally true, there are limitations. For example, assume a 12-bit ADC with 3.3-μs conversion time and a 5-V full-scale range. Such an ADC resolves changes in inputs of less than a millivolt. Signals well below 1 kHz change by this amount in less than 3.3-μs conversion time. As a simplified-design guideline, a T/H function is required for input frequencies above about 10 Hz, if 12-bit performance is to be maintained in an ADC with 3- to 3.5-μs conversion time.

Another advantage of an ADC with T/H is that there are *no transient loads on the signal source.* ADCs without T/H (especially high-speed ADCs) affect the signal lines. If the signal source does not settle in time for the bit decision, the ADC comparator bases its decision on false information, resulting in converter noise and nonlinearity at best, and possible inaccuracy at worst. This is true even for low-frequency signals that do not violate the ½ LSB rule and causes many successive-approximation ADCs (and even some flash ADCs) to place high demands on the signal source, particularly with 12 bits and more.

A sampling ADC converter (with mux and T/H as shown in Fig. 3-2) demands very little from the signal source in terms of dynamic capability, even though the ADC is of the successive-approximation type. Current is drawn from the input source only once when the input is sampled, not on each bit test. This accommodates a much wider variety of signal sources without buffering.

3.5 Interfacing

The ideal data-converter interface is fast, requires a minimum of hardware (preferably the existing buses and control lines) and existing software (available microprocessor control signals). Unfortunately, the specific converter interface architecture that accomplishes this varies widely with application. Common microprocessors use 8- and 16-bit buses. Microcontrollers use 4- and 8-bit buses. Both can use serial interfaces.

Figure 3-3 shows the timing diagram of the control interface between a typical ADC (the MAX163/4/7) and a microprocessor. This particular ADC connects to either 8- or 16-bit buses, and has selectable interface modes using only one or two control lines. In one configuration, the microprocessor can start a conversion, perform other tasks, and then come back to read data. In the other mode, the microprocessor starts with a read ($\overline{\text{RD}}$), and waits until the ADC supplies the answer. The microprocessor uses a $\overline{\text{BUSY}}$ output on the ADC to stretch the read cycle. This makes the ADC appear like a slow-memory device to the system.

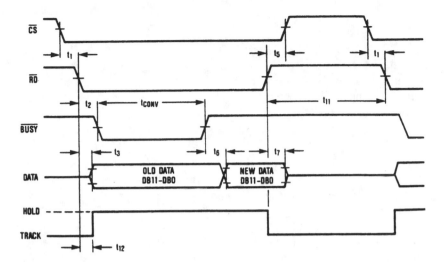

FIGURE 3-3 Timing diagram of control interface between ADC and microprocessor (*Maxim New Releases Data Book*, 1992, p. 7-13)

3.5.1 Serial Interface

Serial interfaces can provide major advantages, especially in data-acquisition designs, in which size and board space are often at a premium. Consider the space requirements of two equivalent 12-bit ADCs. A serial device (such as the MAX170) fits into an 8-pin package, whereas a parallel ADC (the MAX162/172) uses 24 pins. In addition, because of the narrow data path (typically no more than three lines) the serial device does not affect conversion time. Serial interfaces should also be considered when designing electrically isolated systems (see Section 3.5.2).

3.5.2 Electrical Isolation with Opto-Coupling

Some operating environments do not allow electrical connection between the signal source and the measuring system. This is a frequent safety requirement in industrial applications in which it is possible for the input signals to be "hot" in reference to the AC power line, either unintentionally or by design. The most common method for electrical isolation in data-conversion systems is opto-coupling. Figure 3-4 shows an ADC with full electrical isolation in which three opto-coupled lines are used (serial data, clock, and conversion start).

With opto-couplers to isolate data lines, the ADC transmits information, but without direct electrical connection to the microprocessor (or other system device). Both the ADC and analog input signal are isolated from the processor and system. A serial ADC works best because only the three interface lines require opto-couplers.

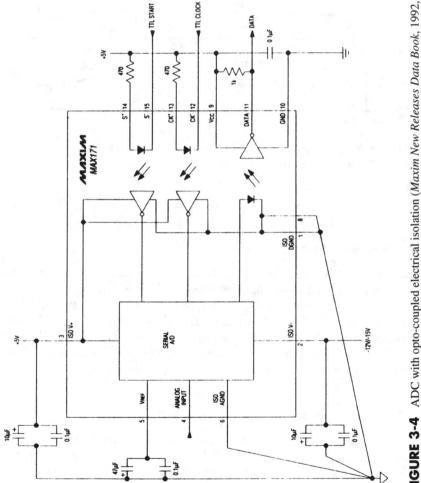

FIGURE 3-4 ADC with opto-coupled electrical isolation (*Maxim New Releases Data Book*, 1992, p. 7-14)

Isolation provided by the ADC of Fig. 3-4 also has the advantage of noise reduction by eliminating ground loops and by providing superior common-mode rejection to most differential-input devices. This is especially useful in large systems in which signals from many remote locations are returned to one point.

Each signal return line is likely to be at a slightly different potential, so ground-loop currents are almost certain.

3.6 Conversion Speed

Some ADC manufacturers specify operating speed in terms of *conversion time* (the time required for one conversion). Other manufacturers use *conversions* (or samples) *per second*. Although these two terms are closely related, the *true conversion rate* does not always translate exactly to the inverse of conversion time. This is because ADCs often require time between conversions.

Figure 3-5 shows the timing diagram of a typical ADC (the MAX167) that performs a conversion in 7.81 µs (12.5 clock cycles), if the start of the conversion is synchronous with a 1.6-MHz clock. Data-access timing limits prevent the next conversion from starting immediately. However, keeping the times as short as possible, and guaranteeing data-bus timing specifications over a given temperature range, minimizes dead time between conversions. This allows the ADC continuously to perform 100,000 samples per second, including the operating time of the internal T/H circuits.

The ADC of Fig. 3-5 can "pipe line" the output data bits to increase interface speed. This means that as each new conversion is started by a signal on $\overline{RD}$ (read), the results of the last conversion (the digital bits representing the analog input) immediately appear on the data bus. The data bits are always one conversion old but can be accessed without waiting. Separate read and write commands are not needed.

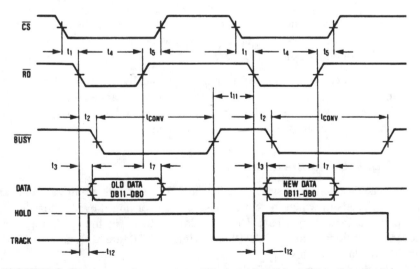

FIGURE 3-5 Timing diagram of parallel-read ADC (*Maxim New Releases Data Book*, 1992, p. 7-15)

3.7 AC Signal Processing

When an ADC measures an AC signal, deviations from the ideal-converter transfer function generate noise and distortion in the digital-output data. The severity of these distortions might not correlate with classic ADC specifications, such as offset, linearity, and full-scale error. In general, large offset and full-scale errors do little harm to AC performance. The effect on linearity usually depends more on the shape of the error rather than the amplitude. From a simplified-design standpoint, converter noise expressed as a portion of an LSB does not indicate the effect on dynamic signals. What is important for DC-signal measurements is not necessarily important for AC signals.

3.8 Noise Problems

Noise pickup can be a problem when converting any analog signal to digital, but it is particularly troublesome when the analog-signal amplitudes are small (about 100 mV or less). As a simplified-design guideline, make the *ADC conversion as close to the source as practical.* (This is easily done with small, serial-output ADCs such as the MAX170.) Another simplified-design technique is to *amplify the analog inputs* (again near the source) to the highest practical level. In any ADC or DAC system, try to minimize noise pickup by *routing any low-level signals away* from clock lines, relays, and any other noisy switching circuits.

3.8.1 Nodes Particularly Susceptible to Noise

The following circuit nodes or points often have noise problems:

1. ADC input pins not driven from a buffered source
2. The investing inputs of op amps
3. The center nodes on signal-divider networks
4. The outputs of unbuffered filters

These nodes or points should be made as physically small as practical to reduce coupling from noise sources. Analog circuits should be separated from digital circuits on the board whenever practical. In the case of very low-level analog signals, separately shielded subcircuits should be used if possible.

3.8.2 Using Buffers

Some data converter systems (especially ADCs) might require buffers, particularly in circuits in which there are missing codes or noise, even when the source and physical layout are clean. The problem might be at the ADC input. Many successive-approximation ADCs (those without T/H) are difficult to drive and should have input buffers.

Use the following simplified-design test to determine if buffers are needed (or would greatly improve performance). Slow the system clock down to about half speed. This allows the input source more time to settle after being loaded by the ADC input circuits. If a slower clock cures the noise problem (restores any missing codes), the input signal probably needs a buffer. Try the circuit with a buffer at the normal clock speed as a final check.

3.8.3 Noise and the ADC Reference

In many data-acquisition applications, the fact that the ADC reference input also acts as an analog input is often overlooked. This is important because some ADC architectures are less effective than others in rejecting reference noise. For example, if the reference voltage settles poorly in response to current demands in the ADC IC, noise can result. So if you have noise that cannot be cured by any of the other methods described in this section, try an ADC with different architecture in the same circuit.

3.9 Practical Layout

The following tips apply to all of the simplified design examples in the remaining chapters of this book.

3.9.1 Single-Point Grounding

Try to connect all signal, ADC/DAC, and power grounds to a single point. This minimizes ground loops that could generate unwanted currents produced by stray magnetic fields. Single-point grounding also prevents the voltage drop in current-carrying lines from affecting the ground reference of more sensitive circuits.

3.9.2 Separate Power Supplies

If practical, use separate power supplies for the analog and digital circuits. Ideally, analog and digital circuits should be completely isolated from one another, especially for precision measurement applications. If separate supplies are not practical, connect the analog and digital circuits to the single system supply through separate RC filters.

3.9.3 Separate Ground Pins

By their very nature, ADCs and DACs commingle internal digital and analog circuits. However, if a converter IC has both analog and digital ground pins, it is best to connect both to the one best (quietest and usually analog) ground only.

3.9.4 Shield Grounds

When shielded cables are used, ground the shield at one end only, preferably at the end nearest the single-point ground.

CHAPTER **4**

Simplified Design with a Typical ADC

This chapter is devoted to simplified-design approaches for a typical ADC IC. All of the general design information in Chapters 1 through 3 applies to the examples in this chapter. However, each data-converter IC has special design requirements, all of which are discussed in detail. The circuits in this chapter can be used immediately the way they are or, by altering component values, as a basis for simplified design of similar data-converter applications. The chapter concludes with a fully isolated 12-bit ADC using serial-to-parallel conversion.

4.1 General Description of ADC

Figure 4-1 shows the functional block diagram and pin configuration for the ADC (a MAX187/189). This serial-bit IC operates from a single $+5$-V supply and accepts a 0-V to 5-V analog input. The ICs feature an 8.5-μs SAR, a 1.5-μs T/H, an on-chip clock, and a high-speed serial interface. An external clock accesses data from the interface, which communicates without external hardware to most digital processors and microcontrollers. The interface is compatible with SPI™, QSPI™, and Microwire™. The ICs digitize signals at a 75-ksps (75,000 samples per second), with a normal power consumption of 7.5 mW, and a shutdown power of 10 μW. The MAX187 has an on-chip buffered reference. The MAX189 requires an external reference.

4.2 Data-Converter Operation

Figures 4-2 and 4-3 show the ADCs in their simplest configurations. Figure 4-4 shows the pin descriptions. As shown in Figure 4-1, the ICs use input T/H and a 12-bit SAR to convert an analog input signal to a digital 12-bit output, as described in Chapter 1. No external-hold capacitor is needed for the T/H. Conversion of input signals in the 0-V to V_{REF} range are converted in 10 μs, including the T/H acquisition time. The internal reference of the MAX187 is trimmed to 4.096 V. Both ICs can accept external references from $+2.5$ V to V_{DD}.

65

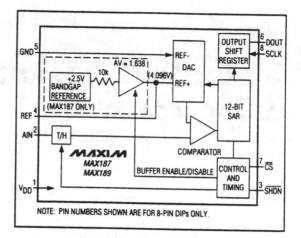

FIGURE 4-1

Functional block diagram and pin configuration for ADC (*Maxim New Releases Data Book*, 1995, p. 7-73). SPI™ and QSPI™ are trademarks of Motorola. Microwire™ is a trademark of National Semiconductor

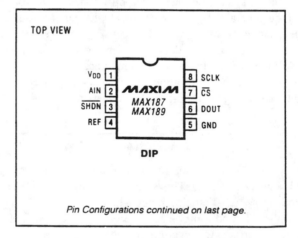

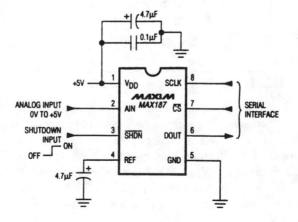

FIGURE 4-2

MAX187 basic operational diagram (*Maxim New Releases Data Book*, 1995, p. 7-51)

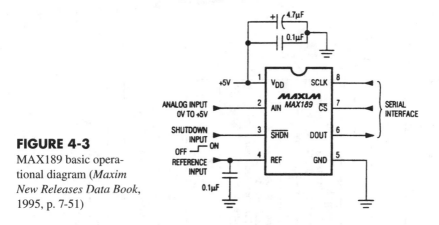

FIGURE 4-3

MAX189 basic operational diagram (*Maxim New Releases Data Book*, 1995, p. 7-51)

The serial interface requires only three digital lines, SCLK, $\overline{CS}$, and DOUT (Fig. 4-4) to provide easy interface to microprocessors. There are two operating modes: normal and shutdown. Pulling the $\overline{SHDN}$ pin low shuts the IC down and reduces supply current to less than 10 µA. Pulling $\overline{SHDN}$ high (or leaving the pin floating) puts the IC in the operating mode. A conversion is initiated by making $\overline{CS}$ low. The conversion result is available at DOUT in unipolar serial format. A high bit that signals end of conversion (EOC) followed by the data bits (MSB first) makes up the serial data stream. The MAX187 internal reference is selected when $\overline{SHDN}$ is formed high. $\overline{SHDN}$ is left floating to select external reference.

4.2.1 Analog Input

Figure 4-5 shows the equivalent analog-input circuits. The full-scale input voltage depends on the voltage at REF, as described in Fig. 4-4. Note that a bypass is recommended at the REF pin for both internal and external reference configurations.

4.2.2 Track and Hold

In the track mode, the analog signal is acquired and stored in the internal hold capacitor (CHOLD) (Fig. 4-5). In hold mode, the T/H switches open and maintains a constant input to the SAR circuits. During acquisition, the analog input (AIN) charges CHOLD. Bringing $\overline{CS}$ low ends the acquisition interval. At this instant, the T/H switches the input side of CHOLD to ground (GND). The retained charge on CHOLD represents a sample of the input, unbalancing the ZERO point at the comparator input.

In the hold mode, the capacitive DAC adjusts during the remainder of the conversion cycle to restore the ZERO point to 0V, within the limits of a 12-bit resolution. This action is equivalent to transferring a charge from CHOLD to the binary-weighted capacitive DAC, which forms a digital representation of the analog input signal. At the end of conversion, the input side of CHOLD switches back to AIN, and CHOLD charges to the input signal again.

PIN		NAME	FUNCTION
DIP	WIDE SO		
1	1	V_{DD}	Supply voltage, +5V ±5%
2	3	AIN	Sampling analog input, 0V to V_{REF} range
3	6	SHDN	Three-level shutdown input. Pulling SHDN low shuts the MAX187/MAX189 down to 10µA (max) supply current. Both MAX187 and MAX189 are fully operational with either SHDN high or floating. For the MAX187, pulling SHDN high enables the internal reference, and letting SHDN float disables the internal reference and allows for the use of an external reference.
4	8	REF	Reference voltage—sets analog voltage range and functions as a 4.096V output for the MAX187 with enabled internal reference. REF also serves as a +2.5V to V_{DD} input for a precision reference for both MAX187 (disabled internal reference) and MAX189. Bypass with 4.7µF if internal reference is used, and with 0.1µF if an external reference is applied.
5	—	GND	Analog and digital ground
—	10	AGND	Analog ground
—	11	DGND	Digital ground
6	12	DOUT	Serial data output. Data changes state at SCLK's falling edge.
7	15	CS	Active-low chip select initiates conversions on the falling edge. When CS is high, DOUT is high impedance.
8	16	SCLK	Serial clock input. Clocks data out with rates up to 5MHz.
—	2,4,5,7,9,13,14	N.C.	Not internally connected. Connect to AGND for best noise performance.

FIGURE 4-4 MAX187/189 pin descriptions (*Maxim New Releases Data Book*, 1995, p. 7-49)

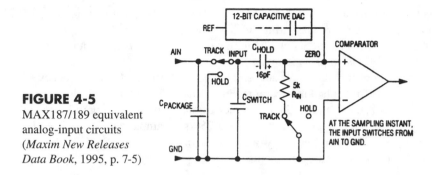

FIGURE 4-5

MAX187/189 equivalent
analog-input circuits
(*Maxim New Releases
Data Book*, 1995, p. 7-5)

In any ADC with T/H at the input, the time required for the T/H to acquire an input signal is a function of how quickly the input capacitance is charged. If the input-signal source impedance is high, the acquisition time is longer, and more time must be allowed between conversions. The acquisition time for these ADCs is calculated as follows: acquisition time $= 9 (R_S + R_{IN})$ 16 pF, where R_{IN} is 5k, R_S is the source impedance of the input signal, and the time is never less than 1.5 µs. *For simplified design,* source impedances less than 5k do not significantly affect the AC performance of these ADCs.

4.2.3 Input Bandwidth

The input-tracking circuit has a 4.5-MHz small-signal bandwidth, and an 8 V/µs slew rate. It is possible to convert (or digitize) high-speed events and measure periodic signals when the bandwidths exceed the sampling rate. This requires under-sampling, however, and is generally not recommended unless an ADC with a higher sampling rate is not practical. One problem with undersampling is aliasing of unwanted high-frequency signals into the frequency band of interest. This can be overcome with an anti-alias filter, but that requires additional components (see the MAX274/MAX275 continuous-time filter data sheet for filter information). *For simplified design,* use an ADC with a higher sampling rate (if possible), such as described in Chapter 5.

4.2.4 Input Protection

The ADCs in this chapter are provided with internal protection diodes that clamp the analog input to swing from GND -0.3 V to V_{DD} $+0.3$ V without damage. However, for accurate conversions near full scale, the input must not exceed V_{DD} by more than 50 mV or be lower than GND by 50 mV. If the analog input exceeds the supplies by more than 50 mV in either direction, limit the input current to 2 mA. Any larger currents can degrade conversion accuracy on these ADCs. There is usually a similar limit for most ADCs.

4.2.5 Analog Input-Drive Amplifiers

The input lines to AIN and GND should be kept as short as possible to mini-mize noise pickup (a classic guideline for all ADCs, and most ICs). Shield any leads that must be long. Because these ADCs have a T/H input, the drive requirements of the op amp driving AIN are less stringent that those in which no T/H is provided. (That is one of the major benefits of T/H inputs in an ADC.) A typical input capacitance for these ADCs is 16 pF.

The input amplifier must have sufficient bandwidth to handle the frequency of the input signal. The manufacturer recommends a MAX400 and 0P07 for lower frequencies and a MAX427 and OP27 for higher frequencies. Keep in mind that the allowed input frequency range is limited by the 75-ksps sample rate. Therefore, the maximum sinusoidal input frequency allowed is one-half or 37.5 kHz. Higher-frequency signals cause aliasing problems.

4.2.6 Internal Reference

The internal reference of the MAX187 is connected to the REF pin and to the internal DAC. This reference can source up to 0.6 mA for use by external components connected to the REF pin. Decouple the REF pin to ground with a 4.7-μF capacitor. The internal reference can be disabled when $\overline{\text{SHDN}}$ is left floating. This allows use of an external reference.

4.2.7 External Reference

If an external reference is used (always for MAX189 and when the MAX187 internal reference is disabled), stay within the voltage range from +2.5 V to V_{DD}. The external reference must be capable of delivering up to 350 μA with an output impedance of 10 ohms or less. Decouple the REF pin to ground with a minimum of 0.1 μF. If the reference has a high output impedance (or is noisy), use 4.7 μF for the bypass.

4.3 Serial Interface Considerations

Figure 4-6 shows the complete conversion sequence of the serial interface lines, including the shutdown sequence. Figures 4-7 and 4-8 show details of the interface timing sequence. The following paragraphs summarize the serial-interface functions.

4.3.1 Initialization and Conversion Start

When power is first applied, it takes the fully discharged 4.7-μF reference-bypass capacitor about 20 ms to provide adequate charge (for specified accuracy). With $\overline{\text{SHDN}}$ not pulled low, the ADCs are now ready to convert.

To start a conversion, pull $\overline{\text{CS}}$ low. At the falling edge of $\overline{\text{CS}}$, the T/H enters the hold mode, and a conversion is initiated. After an internally timed 8.5-μs conversion period, the end of conversion is indicated when DOUT goes high. Data can then be shifted out serially with the external clock.

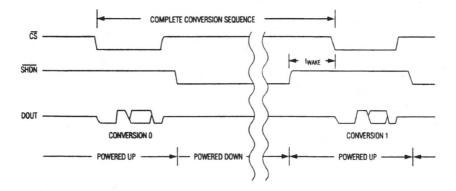

FIGURE 4-6 Complete conversion sequence of serial-interface lines including shutdown (*Maxim New Releases Data Book*, 1995, p. 7-52)

4.3.2 Reducing Supply Current

You can reduce power consumption considerably by shutting down the ICs between conversions. This is shown in Fig. 4-9. Because the MAX189 uses an external reference voltage, it "wakes up" faster than the MAX187 and thus provides lower average supply currents. The wakeup time (tWAKE) is the time from deassertion of $\overline{\text{SHDN}}$ to the time when a conversion can be initiated. For the MAX187, this time is 2 µs. For the MAX189, this time depends on the time in shutdown (Fig. 4-10). This is because the external 4.7-µs reference bypass capacitor loses its charge slowly during shutdown.

4.3.3 External Clock

The actual conversion does not need the external clock. This frees the microprocessor from the burden of running the SAR conversion clock and allows the conversion result to be read back at the convenience of the microprocessor at any clock rate from 0 MHz to 5 MHz. The clock duty cycle is unrestricted if each clock phase is at least 100 ns. *Do not* run the clock while a conversion is in progress.

4.3.4 Timing and Control

As shown in Figs. 4-7 and 4-8, conversion-start and data-read operations are controlled by the $\overline{\text{CS}}$ and SCLK digital inputs. A $\overline{\text{CS}}$ falling edge initiated a conversion sequence: the T/H falling edge holds input voltage, the ADC begins to convert, and DOUT changes from high impedance to logic low. SCLK must be kept inactive during the conversion. An internal register stores the data when the conversion is in progress.

End of conversion (EOC) is indicated by DOUT going high. The DOUT rising edge can be used as a framing signal. SCLK shifts the data out of the register any time

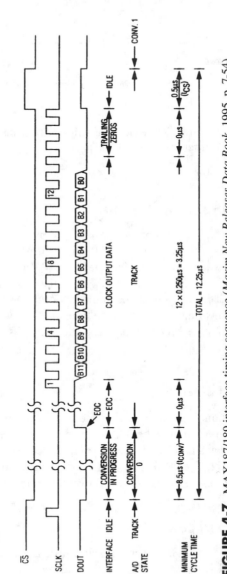

FIGURE 4-7 MAX187/189 interface timing sequence (*Maxim New Releases Data Book*, 1995, p. 7-54)

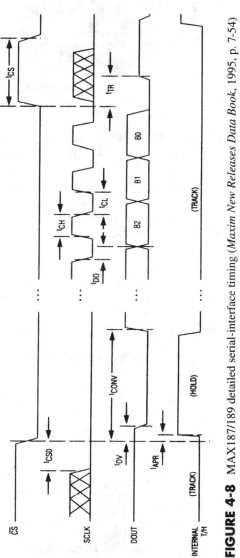

FIGURE 4-8 MAX187/189 detailed serial-interface timing (*Maxim New Releases Data Book*, 1995, p. 7-54)

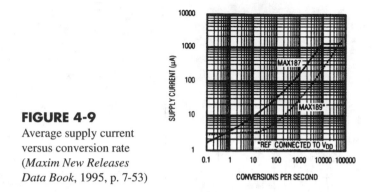

FIGURE 4-9

Average supply current
versus conversion rate
(*Maxim New Releases
Data Book*, 1995, p. 7-53)

after the conversion is complete. The DOUT transitions occur on the SCLK falling
edge. The next falling clock-edge produces the MSB of the conversion at DOUT, followed by the remaining bits.

Because there are 12 data bits and one leading high bit, at least 13 falling clock
edges are needed to shift out a full set of bits. Extra clock pulses after the conversion
result has been clocked out and before a rising edge of $\overline{CS}$ produce trailing 0s at
DOUT and have no effect on converter operation.

To get minimum cycle time, use the DOUT rising edge as the EOC signal. Clock
out the data with 13 clock cycles at full speed, and raise $\overline{CS}$ after the LSB of the conversion has been read. After the specified minimum acquisition time (tACQ), $\overline{CS}$ can
be pulled low again to initiate the next conversion.

4.3.5 Output Coding and Transfer Function

As shown in Fig. 4-11, the data output from these ADCs is binary (unipolar),
and code transitions occur halfway between successive integer LSB values. If V_{REF}
is +4.096 V (the trimmed value for the MAX187), then 1 LSB = 1.00 mV or
4.096V/4096.

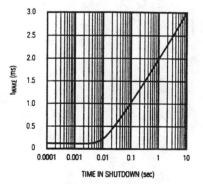

FIGURE 4-10

MAX187 tWAKE versus
time in shutdown (*Maxim
New Releases Data Book*,
1995, p. 7-53)

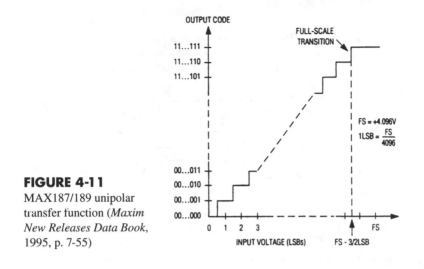

FIGURE 4-11

MAX187/189 unipolar transfer function (*Maxim New Releases Data Book*, 1995, p. 7-55)

4.4 Dynamic Performance Characteristics and Testing

As discussed in Chapter 2, ADCs are often tested for such characteristics as zero error, full-scale error, integral nonlinearity (INL), and differential nonlinearity (DNL). These tests are generally sufficient when the ADC is used to convert DC and slowly varying signals. However, when the ADC is used with dynamic input signals (wideband signal processing), other tests and parameters are of much greater importance. The following paragraphs summarize the most important dynamic tests from a simplified-design standpoint.

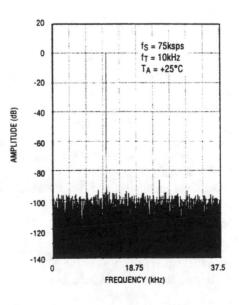

FIGURE 4-12

MAX187/189 FFT plot (*Maxim New Releases Data Book*, 1995, p. 7-55)

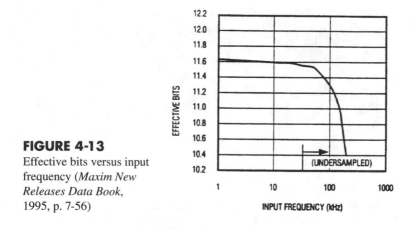

FIGURE 4-13

Effective bits versus input frequency (*Maxim New Releases Data Book*, 1995, p. 7-56)

4.4.1 Fast Fourier Transforms

Figure 4-12 shows a fast fourier transform (FFT) plot using a spectrum analyzer for the MAX187/189. Such tests guarantee the dynamic frequency response, distortion, and noise at the rated throughput (combination of sampling frequency and total or test frequency) of the ADC. The test involves applying a low-frequency sine wave to the ADC input and recording the digital conversion results *for a specified time*. The data bits are then analyzed using an FFT algorithm that determines the spectral content. Conversion errors are seen as spectral elements outside the fundamental input frequency.

4.4.2 SINAD and Effective Number of Bits

SINAD (signal-to-noise plus distortion) is the ratio of the RMS amplitude of the fundamental input frequency to the RMS amplitude of all other ADC output signals. The input bandwidth is limited to frequencies above DC and below one-half of the ADC sample (conversion) rate.

The theoretic minimum ADC noise is caused by quantization error and is a direct result of the resolution of the ADC: SINAD = $(6.02N + 1.76)$ dB, where N is the number of bits of resolution. An ideal 12-bit ADC can therefore do no better than 74 dB. However, as shown in Fig. 4-12, there is an output signal (at about 12 kHz) greater than 80 dB above signals of other frequencies.

The effective number of bits (or effective resolution) that can be provided by an ADC is determined by transposing the SINAD equation and substituting the measured SINAD: N = $(SINAD - 1.76)/(6.02)$. Figure 4-13 shows the effective number of bits as a function of input frequency for the MAX187/189.

4.4.3 Total Harmonic Distortion

If a pure sine wave is sampled by an ADC at greater than the Nyquist frequency, the nonlinearities in the transfer function of the ADC produce harmonics of the input frequency present in the sampled output data. For an ADC, total harmonic distortion

(THD) is the ratio of the RMS sum of all harmonics (in the frequency band above DC and below one-half the sample rate, but not including the DC component) to the RMS amplitude of the fundamental frequency. This is expressed as:

$$\frac{V2^2 + V3^2 + V4^2 + \ldots VN^2}{V1}$$

where V1 is the fundamental RMS amplitude, and V2 through VN are the amplitudes of the second through Nth harmonics. (The THD specification for the MAX187/189 is -80 dB maximum, with a 10-kHz sine wave, from 0 V to 4.096 V_{p-p}, sampled at 75 ksps).

4.4.4 Using the Dynamic Characteristics

For simplified design, it is not usually necessary to test all the dynamic characteristics. Simply use the ADC data-sheet characteristics. For example, if you are

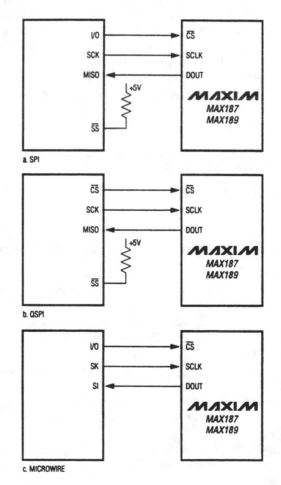

FIGURE 4-14
MAX187/189 interface to three common configurations (*Maxim New Releases Data Book*, 1995, p. 7-56)

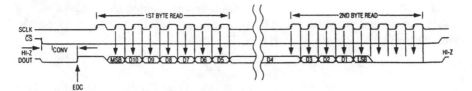

FIGURE 4-15 Serial interface timing for SPI™ or Microwire™ (*Maxim New Releases Data Book*, 1995, p. 7-57)

concerned with accuracy, the relative accuracy (a DC specification) for the top of the line (MAX187A/MAX189B) is $\pm\frac{1}{2}$ LSB. Figure 4-13 shows that the effective bits (or effective resolution) is within $\frac{1}{2}$ LSB at input frequencies up to and more than 20 kHz. Therefore it is reasonable to assume that both accuracy and resolution are within the $\frac{1}{2}$-bit limit over the same input-frequency range.

One exception to this simplified rule is a situation in which the input frequency is near the test limits or the rated limits of the ADC. Should that occur, you can test the ADC in a basic circuit (Figs. 4-2 and 4-3), as described in Section 1.4.9, at the maximum input frequency. If the ADC performs properly, but you are still unsure, increase the input frequency to well beyond anticipated value (say at 1.5 or 2 times the maximum), and repeat the accuracy tests. If the ADC does not provide the required accuracy at (or slightly above) the maximum frequency, it may be necessary to use another ADC (probably at greater expense). Chapter 5 describes ADCs suitable for high-frequency input signals.

4.5 Applications Data

The remainder of this chapter describes how the MAX187/189 can be used for specific applications.

4.5.1 Standard Interfaces (General Instructions)

Figure 4-14 shows the basic connections for interface between the MAX187/189 and the three common or standard serial-interface configurations. Keep the following in mind when using any of the three interfaces.

Set the microprocessor or CPU (central-processing unit) serial interface in the *master mode* so that the CPU will generate the serial clock. Choose a clock frequency up to 2.5 MHz.

Use a general-purpose I/O (input-output) line on the CPU to pull $\overline{CS}$ low. Keep SCLK low.

Wait for the maximum conversion time (8.5 μs) specified before activating SCLK. As an alternative, look for a DOUT rising edge to determine the end of conversion.

Activate SCLK for a minimum of 13 clock cycles. The first falling clock edge will produce the MSB of the DOUT conversion. The transition for DOUT data occurs on the SCLK falling edge and is available in MSB-first format. Observe the SCLK to DOUT valid-timing characteristics (Figs. 4-6 through 4-8). Data can be clocked into the microprocessor on the SCLK rising edge.

Pull $\overline{CS}$ high at or after the 13th falling clock-edge. If $\overline{CS}$ remains low, trailing zeros are clocked out after the LSB.

With $\overline{CS}$ high, wait the minimum specified time (tCS) before launching a new conversion by pulling $\overline{CS}$ low. (The minimum tCS is 500 ns for these ADCs.) If a conversion is aborted by pulling $\overline{CS}$ high before the conversion ends, wait for the minimum acquisition time (tACQ) before starting a new conversion. (The minimum tACQ is 1.5 µs.)

The data bits can be output in 1-byte chunks or continuously (Fig. 4.7). The bytes contain the result of the conversion, padded with one leading 1, and trailing 0s, if SCLK is still active with $\overline{CS}$ kept low.

4.5.2 SPI™ and Microwire™ Interface

Figure 4-15 shows the serial interface timing when the MAX187/189 is used with SPI™ or Microwire™. (Note that with SPI™, it is necessary to set the interface CPOL and CPHA lines to 0.)

Conversion starts with a $\overline{CS}$ falling edge. DOUT goes low, indicating a conversion in progress. Wait until DOUT goes high (or the maximum specified 8.5 µs conversion time). Two consecutive 1-byte reads are required to obtain the full 12 bits. Output-data (DOUT) transitions occur on the SCLK falling edge and are clocked into the microprocessor on the SCLK rising edge. As shown in Fig. 4-15, the first byte contains a leading 1 as well as 7 bits of conversion. The second byte contains the remaining 5 bits and three trailing zeros.

4.5.3 QSPI™ Interface

Figure 4-16 shows the serial interface timing when the MAX187/189 is used with QSPI™. (Again, it is necessary to set the interface CPOL and CPHA lines to 0.) Unlike SPI™, which requires two 1-byte reads to acquire the 12 bits of data from the ADC, QSPI™ allows the minimum number of clock cycles necessary to clock the data. The ADCs require 13 clock cycles from the microprocessor to clock out the 12 bits of data with no trailing 0s. The minimum clock frequency to ensure compatibility with QSPI™ is 2.77 MHz.

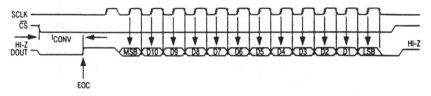

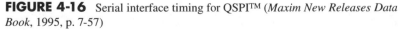

FIGURE 4-16 Serial interface timing for QSPI™ (*Maxim New Releases Data Book*, 1995, p. 7-57)

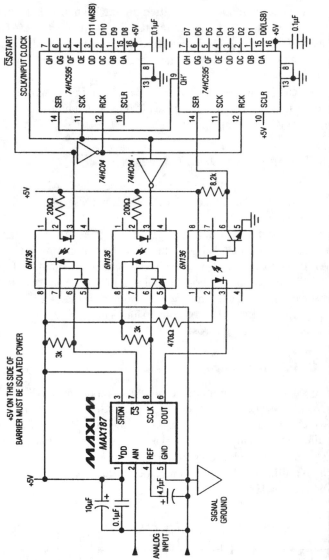

FIGURE 4-17 12-bit fully isolated ADC with serial-to-parallel conversion (*Maxim New Releases Data Book*, 1995, p. 7-58)

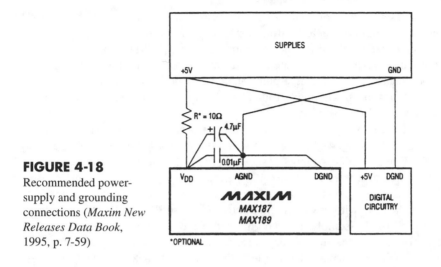

FIGURE 4-18
Recommended power-
supply and grounding
connections (*Maxim New
Releases Data Book*,
1995, p. 7-59)

4.5.4 Fully Isolated Serial-to-Parallel Conversion

Figure 4-17 shows the MAX187 connected to provide full 12-bit ADC operation with serial-to-parallel conversion. This circuit operates from a single 5-V supply
and is fully isolated using opto-couplers. The circuit is well suited to industrial applications in which the control electronics must be separated or isolated from hazardous
electrical conditions. The circuit also provides noise immunity and prevents excessive current flow where there are different ground levels between the ADC and the
rest of the system. The circuit does not require expensive isolation amplifiers and is
cost effective because the opto-isolation is in the serial link (between the ADC and
serial-parallel converters).

The ADC results are transmitted across a 1,500-V isolation barrier provided
by three 6N136 opto-isolators. (Isolated power must be supplied to the ADC and
the isolated side of the opto-couplers.) The 74HC595 three-state shift registers are
used to convert the serial-data bits into a 12-bit parallel-data output. Conversion
speed is limited by the delay through the opto-isolators. With a 140-kHz clock, conversion time is 100 μs. The circuit also can be used without the opto-couplers if
electrical isolation is not required. When the opto-couplers are eliminated, the clock
can increased up to 2.9 MHz without violating the 20-ns set-up time required by the
shift registers.

4.5.5 Practical Considerations for the MAX187/189

Figure 4-18 shows the recommended power-supply and grounding connections. The following points should be considered when one uses the MAX187/189,
or any other similar ADC.

Use the single-point or "star" grounding system shown in Fig. 4-18 at the GND terminal. This should be separate from the logic ground. All other analog grounds should be connected to the single-point ground. Connect DGND to the star or single-point ground for further noise reduction. However, no other digital-system ground should be connected to the single-point analog ground. The ground return to the power supply for the single-point ground should be low impedance and as short as possible for noise-free operation.

High-frequency noise in the V_{DD} power supply can affect the high-speed comparator in the ADC (Fig. 4-1). Bypass this supply to the single-point analog ground with 0.01-µF and 4.7-µF capacitors as shown in Fig. 4-18. Keep capacitor leads short for best supply-noise rejection. If the +5-V supply is very noisy, a 10-ohm resistor can be connected as a lowpass filter to attenuate supply noise as shown in Fig. 4-18.

As is the case with any ADC (and most ICs), use PC boards, not wire-wrap boards, for all final assembly. Keep digital and analog signal lines separated from each other (whenever possible). In any case, do not run analog and digital (especially clock) lines parallel to one another or run digital lines underneath the ADC package.

CHAPTER **5**

Simplified Design with a Flash ADC

This chapter is devoted to simplified-design approaches for a typical video-frequency (VF) or high-speed flash-type ADC. All of the general design information in Chapters 1 through 3 applies to the examples in this chapter. The circuits in this chapter can be used immediately the way they are or, with alterations in component values, as a basis for simplified design of similar data-converter applications. The chapter concludes with a high-speed interface circuit.

5.1 General Description of ADC

Figure 5-1 shows the functional block diagram for the ADC (a Raytheon Semiconductor TDC1147). The pin configuration is shown in Fig. 5-2. The package interconnections are shown in Fig. 5-3.

This 7-bit flash ADC has no "pipeline delay" between sampling and valid data. The output-data register normally found on flash ADCs is bypassed, allowing data to transfer directly to output drivers from the encoding-logic section of the circuit. (The TDC1147 is function- and pin-compatible with the Raytheon Semiconductor TDC1047, which does have an output-data register.) The TDC1147 requires only one clock pulse to perform the complete conversion operation. Conversion time is guaranteed to be less than 60 ns. The TDC1147 operates accurately at sampling rates up to 15 Msps (15 million samples per second), and has an analog bandwidth of 7 MHz, as shown in Fig. 5-4. Linearity errors are guaranteed to be less than 0.4% over the operating temperature range.

The TDC1147 is fully TTL compatible, as shown in Figs. 5-3 and 5-5, and has 7-bit resolution with ½ LSB linearity. The output coding (Fig. 5-6) format can be selected. A sample-and-hold circuit is not required for any configuration.

5.2 Data-Converter Operation

As shown in Fig. 5-1, the TDC1147 has two functional sections: a comparator array and encoding logic. The comparator array compares the input signal with 127 reference voltages to produce an N-of-127 code. (Such a code is sometimes called the

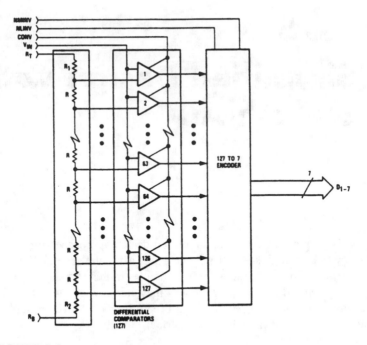

FIGURE 5-1 Functional block diagram of TDC1147 (*Raytheon Semiconductor Data Book*, 1994, p. 3-95)

thermometer code because all comparators referred to voltages more positive than the input signal will be off, and those referred to voltages more negative than the input will be on.) The encoding logic converts the N-of-127 code into binary or offset two's complement and can invert either output code. The coding function is controlled by DC signals on pins NM_{INV} and NL_{INV}.

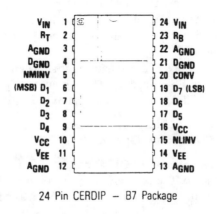

FIGURE 5-2 Pin configuration for TDC1147 (*Raytheon Semiconductor Data Book*, 1994, p. 3-96)

V_{IN}	1	24	V_{IN}
R_T	2	23	R_B
A_{GND}	3	22	A_{GND}
D_{GND}	4	21	D_{GND}
NMINV	5	20	CONV
(MSB) D_1	6	19	D_7 (LSB)
D_2	7	18	D_6
D_3	8	17	D_5
D_4	9	16	V_{CC}
V_{CC}	10	15	NLINV
V_{EE}	11	14	V_{EE}
A_{GND}	12	13	A_{GND}

24 Pin CERDIP – B7 Package

Signal Type	Signal Name	Function	Value	B7 Package Pins
Power	V_{CC}	Positive Supply Voltage	+5.0V	10, 16
	V_{EE}	Negative Supply Voltage	−5.2V	11, 14
	D_{GND}	Digital Ground	0.0V	4, 21
	A_{GND}	Analog Ground	0.0V	3, 12, 13, 22
Reference	R_T	Reference Resistor (Top)	0.00V	2
	R_B	Reference Resistor (Bottom)	−1.00V	23
Controls	NMINV	Not Most Significant Bit INVert	TTL	5
	NLINV	Not Least Significant Bit INVert	TTL	15
Convert	CONV	Convert	TTL	20
Analog Input	V_{IN}	Analog Signal Input	0V to −1V	1, 24
Outputs	D_1	MSB Output	TTL	6
	D_2		TTL	7
	D_3		TTL	8
	D_4		TTL	9
	D_5		TTL	17
	D_6		TTL	18
	D_7	LSB Output	TTL	19

FIGURE 5-3   Package interconnections for TDC1147 (*Raytheon Semiconductor Data Book*, 1994, p. 3-97)

5.3 Power Requirements

The TDC1147 operates from two supply voltages, +5.0 V and −5.2 V. The return path for ICC (the current drawn from the +5.0-V supply) is DGND. The return path for IEE (the current drawn from the −5.2-V supply) is AGND. *All* power and ground pins *must be connected.*

5.4 Reference Voltage Requirements

The TDC1147 requires an external reference voltage applied at the input reference-resistor chain. The range of analog input voltages to be converted into digital form is set by the external reference voltages. (V_{RB} is the voltage at the bottom of the chain, and V_{RT} is the voltage at the top of the chain.) The voltage applied across the reference resistor chain ($V_{RT} − V_{RB}$) must be between 0.8 V and 1.2 V. V_{RB} and V_{RT} should be between +0.1 V and −1.1 V. V_{RT} should be more positive than V_{RB} within the range. The nominal voltages are $V_{RT} = 0.00$ V and $V_{RB} = −1.00$ V.

The reference voltages can be varied dynamically up to 7 MHz. Because of slight variations in the reference current with clock and input signals, the RT and RB (top and bottom of reference input chain) points should be low impedance. For circuits in which the reference is not varied, a bypass capacitor to ground is recommended. If the reference inputs are varied dynamically, such as with an AGC (automatic gain control) function, use a low-impedance reference-voltage source.

Parameter		Test Conditions	Temperature Range				Units
			Standard		Extended		
			Min	Max	Min	Max	
E_{LI}	Linearity Error, Integral Independent	V_{RT}, V_{RB} = Nom		0.4		0.4	%
E_{LD}	Linearity Error, Differential			0.4		0.4	%
CS	Code Size	V_{RT}, V_{RB} = Nom	30	170	30	170	% Nominal
V_{OT}	Offset Voltage, Top	$V_{IN} = V_{RT}$		+50		+50	mV
V_{OB}	Offset Voltage, Bottom	$V_{IN} = V_{RB}$		-30		-30	mV
T_{CO}	Temperature Coefficient			±20		±20	µV/°C
BW	Bandwidth, Full Power Input		7		7		MHz
t_{TR}	Transient Response, Full Scale			10		10	ns
SNR	Signal – to – Noise Ratio	7MHz Bandwidth, 20MSPS Conversion Rate					
	Peak Signal/RMS Noise	1MHz Input	45		46		dB
		7MHz Input	43		44		dB
	RMS Signal/RMS Noise	1MHz Input	36		37		dB
		7MHz Input	34		35		dB
E_{AP}	Aperture Error			50		50	ps
DP	Differential Phase Error[1]	F_S = 4 x NTSC		1.5		1.5	Degree
DG	Differential Gain Error[1]	F_S = 4 x NTSC		2.5		2.5	%

Note: 1 In excess of quantization.

FIGURE 5-4 Performance characteristics for TDC1147 (*Raytheon Semiconductor Data Book, 1994,* p. 3-101)

Parameter	Test Conditions	Temperature Range						Units
		Standard			Extended			
		Min	Nom	Max	Min	Nom	Max	
V_{CC}	Positive Supply Voltage (measured to D_{GND})	4.75	5.0	5.25	4.5	5.0	5.5	V
V_{EE}	Negative Supply Voltage (measured to A_{GND})	-4.9	-5.2	-5.5	-4.9	-5.2	-5.5	V
V_{AGND}	Analog Ground Voltage (measured to D_{GND})	-0.1	0.0	0.1	-0.1	0.0	0.1	V
t_{PWL}	CONV Pulse Width, LOW	22			22			ns
t_{PWH}	CONV Pulse Width, HIGH	18			18			ns
V_{IL}	Input Voltage, Logic LOW			0.8			0.8	V
V_{IH}	Input Voltage, Logic HIGH	2.0			2.0			V
I_{OL}	Output Current, Logic LOW			4.0			2.0	mA
I_{OH}	Output Current, Logic HIGH			-0.4			-0.4	mA
V_{RT}	Most Positive Reference Input [1]	-0.1	0.0	0.1	-0.1	0.0	0.1	V
V_{RB}	Most Negative Reference Input [1]	-0.9	-1.0	-1.1	-0.9	-1.0	-1.1	V
$V_{RT}-V_{RB}$	Voltage Reference Differential	0.8	1.0	1.2	0.8	1.0	1.2	V
V_{IN}	Input Voltage	V_{RB}		V_{RT}	V_{RB}		V_{RT}	V
T_A	Ambient Temperature, Still Air	0		70				°C
T_C	Case Temperature				-55		125	°C

Note   1 V_{RT} must be more positive than V_{RB}, and voltage reference differential must be within specified range.

FIGURE 5-5 Operating conditions for TDC1147 (*Raytheon Semiconductor Data Book*, 1994, p. 3-99)

	Binary		Offset Two's Complement	
Range	True	Inverted	True	Inverted
− 1.00V FS	NMINV − 1	0	0	1
	NLINV − 1	0	1	0
0.0000V	0000000	1111111	1000000	0111111
− 0.0078V	0000001	1111110	1000001	0111110
•	•	•	•	•
•	•	•	•	•
•	•	•	•	•
− 0.4960V	0111111	1000000	1111111	0000000
− 0.5039V	1000000	0111111	0000000	1111111
•	•	•	•	•
•	•	•	•	•
•	•	•	•	•
− 0.9921V	1111110	0000001	0111110	1000001
− 1.0000V	1111111	0000000	0111111	1000000

Note

1 Voltages are code midpoints

FIGURE 5-6 Output coding for TDC1147 (*Raytheon Semiconductor Data Book*, 1994, p. 3-101)

5.5 Output Coding Control

Two function-control pins, NM_{INV} and NL_{INV}, are provided to control output coding. These controls require DC (steady state) signals and allow the output coding to be either straight binary or offset two's complement, in either true or inverted sense (Fig. 5-6).

5.6 Initiating a Conversion

The TDC1147 uses a convert (CONV) input signal to initiate the ADC process. Unlike other flash ADCs, which have a one-clock-cycle pipeline delay between sampling and output-data, the TDC requires only a single pulse to perform the entire conversion operation.

As shown in Fig. 5-7, the analog input is sampled (the comparators are latched) within the maximum sampling time offset (tSTO). Data bits from that sample become valid after a maximum output delay time (tD). Data bits from the previous sample are held at the outputs from a minimum output hold time (tHO). This allows data bits from the TDC1147 to be acquired by an external register or other circuit. There are minimum time requirements for the HIGH and LOW portions (tPWH, tPWL) of the CONV waveform, and all output-timing specifications are measured with respect to the rising edge of CONV (Fig. 5-7).

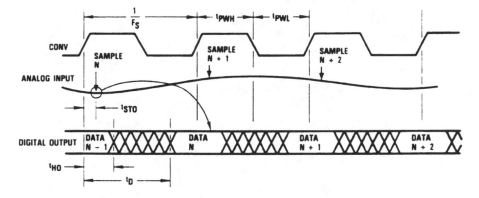

FIGURE 5-7 Timing diagram for TDC1147 (*Raytheon Semiconductor Data Book*, 1994, p. 3-98)

5.7 Analog Inputs

The TDC1147 uses latching comparators that cause the input impedance to vary slightly with the signal level. For best results, both V_{IN} pins must be used, and the source impedance of the driving signal (analog input) must be less than 30 ohms.

Input signals will not damage the TDC1147 if the signals remain within the range of V_{EE} (typically -5.2 V) and $+0.5$ V. If the input signal is between the V_{RT} and V_{RB} references, the output will be a binary number between 0 and 127 inclusive. An input signal outside this range indicates either full-scale positive or full-scale negative, depending on whether the signal is off-scale in the positive or negative direction.

5.8 Digital Outputs

The TDC1147 outputs are TTL compatible and capable of driving four low-power Schottky TTL (54/74 LS) unit loads. The outputs hold the previous data bits a minimum time (tHO) after the rising edge of the CONV signal. New data bits become valid after a maximum time (tD) after the rising edge of the CONV signal. The use of 2.2-k pull-up resistors is recommended at the digital outputs.

5.9 Calibration

Basic calibration of the TDC1147 requires that the voltages applied to the resistor chain, V_{RT} and V_{RB}, be adjusted to set the 1st and 127th thresholds. For example, assuming a 0-V to -1-V input range, apply a -0.0039-V (½ LSB from 0 V) at the analog input and adjust V_{RT} for a digital output that toggles between 00 and 01. Then

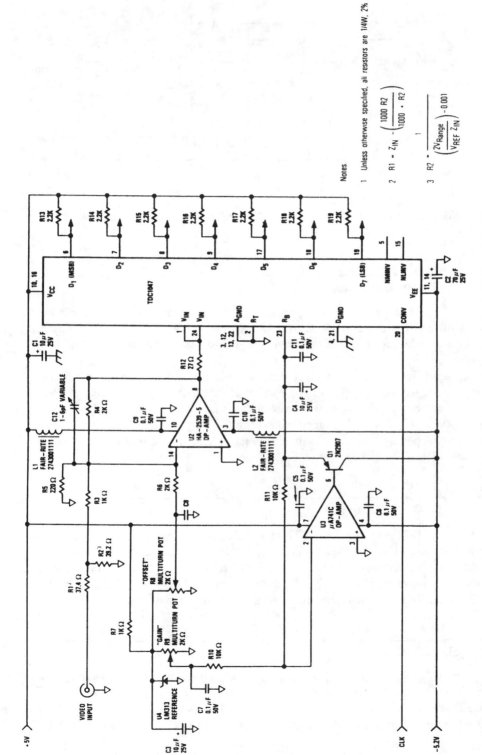

FIGURE 5-8 Video-frequency interface for TDC1147 (*Raytheon Semiconductor Data Book*, 1994, p. 3-103)

Notes

1 Unless otherwise specified, all resistors are 1/4W, 2%

2 $R1 = Z_{IN} - \left(\dfrac{1000 \ R2}{1000 + R2}\right)$

3 $R2 = \dfrac{1}{\left(\dfrac{2V_{Range}}{V_{REF} \ Z_{IN}}\right) - 0.001}$

apply -0.996 V ($\frac{1}{2}$ LSB from -1-V) and adjust V_{RB} for toggling between codes 126 and 127.

This method of calibration requires that both ends of the resistor chain, RT and RB, be driven by variable voltage sources. Instead of adjusting V_{RT}, RT can be connected to the analog ground, and the 0-V end of the range can be calibrated with an input-amplifier offset control. (Such a configuration is described in Section 5.10.) When this alternative calibration is used, an offset error at the bottom of the resistor chain causes a slight gain error, which can be compensated for by varying V_{RB}. The bottom reference is a convenient point for gain adjust that is not in the analog signal path.

The offset voltages are generated by the parasitic resistance between the package pin and the actual resistor chain on the IC. These parasitic resistors are shown as R1 and R2 in Fig. 5-1. Calibration cancels all offset voltage, eliminating any offset and gain errors.

5.10 Typical Video Interface

Figure 5-8 shows a VF interface circuit for the TDC1147. (Note that a TDC1047 is shown. However, the TDC1147 and TDC1047 are pin compatible.) The circuit can be used to convert VF analog signals up to the limits defined in Fig. 5-4. Any of the digital-output formats shown in Fig. 5-6 can be selected with the proper control signals applied to pins 5 and 15 (NM_{INV} and NL_{INV}).

The analog amplifier U2 is a bipolar wideband op amp used to drive the ADC. Bipolar inputs can be accommodated by adjusting the OFFSET control R8. Zener U4 provides a stable reference for R8 as well as the GAIN control R9. U2 has a gain of -1, providing the recommended 1-V_{p-p} input for the ADC.

Variable capacitor C12 allows the frequency response of U2 to be adjusted for optimum performance, depending on the type of analog signal being converted (sine wave, step). C12 can be replaced with a fixed capacitor, if desired, after the circuit-board layout is set, and an optimum value (between 1 and 6 pF) is selected.

The circuit of Fig. 5-8 can be calibrated using the basic techniques described in Section 5.9. Use OFFSET control R8 to set V_{RT}, and GAIN control R9 to set V_{RB}. Note that V_{RB} is supplied by inverting amplifier U3 and transistor Q1. (The PNP transistor is used to provide a low-impedance source and is necessary to sink the current flowing through the reference-resistor chain.) After the 1st and 127th thresholds are set, adjust C12 as necessary for best performance (to accommodate the type of analog signal being converted).

Note that the degree of decoupling shown in Fig. 5-8 might not be required in all configurations. However, proper decoupling is recommended for all power supplies.

CHAPTER 6

Simplified Design with Serial-Interface DAC

This chapter is devoted to simplified-design approaches for a typical DAC with three-wire serial interface. All the general design information in Chapters 1 through 3 applies to the examples in this chapter. The circuits in this chapter can be used immediately the way they are or, with alterations in component values, as a basis for simplified design of similar data-converter applications. The chapter concludes with a typical bipolar output circuit.

6.1 General Description of DAC

Figure 6-1 shows the functional block diagram and pin configuration for a DAC (a MAX512/513). This IC contains three 8-bit, voltage-output DACs (DAC A, DAC B, DAC C). Output buffer-amplifiers for DAC A and DAC B provide voltage outputs and are included in the IC to reduce external component count. The output buffer for DAC A can source or sink 5 mA to within 0.5 V of V_{DD} or V_{SS}. DAC B can can source or sink 0.5 mA to within 0.5 V of V_{DD} or V_{SS}. DAC C is unbuffered, providing a third voltage output with increased accuracy. The MAX512 operates with a single +5 V $\pm$ 10% supply, and the MAX513 operates with a +2.7-V to +3.6-V supply. Both DACs can also operate with split supplies.

The DACs are well suited for portable and battery-operated applications because of ultra-low power consumption and the DIP/SO IC package. Operating supply current is 1 mA, dropping to less than 1 μA at shutdown. Any of the three DACs can be independently shut down. In the shutdown mode, the reference-resistor ladder network of the DAC is disconnected from the reference input, minimizing system power consumption.

The three-wire serial interface has a maximum operating frequency of 5 MHz and is compatible with SPI™, QSPI™, and Microwire™. The serial-input shift register is 16 bits long and consists of 8 bits of DAC input data and 8 bits for DAC selection and shutdown. The DAC registers can be loaded independently or in parallel at the positive edge of $\overline{CS}$. A latched-logic output is also available for auxiliary control.

SPI™ and QSPI™ are trademarks of Motorola. Microwire™ is a trademark of National Semiconductor.

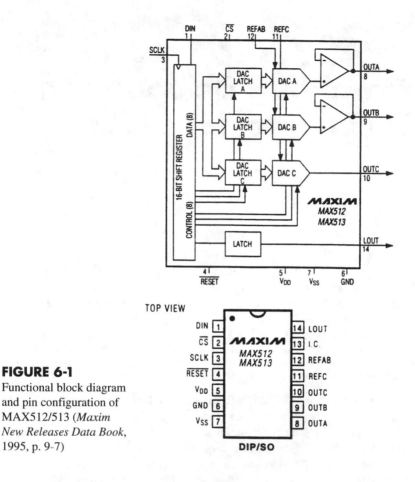

FIGURE 6-1

Functional block diagram
and pin configuration of
MAX512/513 (*Maxim
New Releases Data Book*,
1995, p. 9-7)

6.2 Data-Converter Operation

Figures 6-2 and 6-3 show the pin descriptions and simplified circuit diagram, respectively. The DACs are "inverted" $R-2R$ ladder networks using complementary switches that convert 8-bit digital inputs into equivalent analog output voltages in proportion to the applied reference voltages. (Compare Fig. 6-3 with the circuits of Figs. 1-1 and 1-2.)

The DACs have two reference inputs (see Fig. 6-1). One reference (REF_{AB}) is shared by DAC A and DAC B. The REF_C reference is used only by DAC C. These reference inputs allow different full-scale output voltages, and different output polarities, for the DAC pair A/B and DAC C. All three DACs within the IC operate in either single-supply or dual-supply modes, as determined by V_{SS}. If V_{SS} is within approximately -0.5 V of GND, single-supply mode is assumed. If V_{SS} is less than -1.5 V, the DACs are in dual-supply mode.

PIN	NAME	FUNCTION
1	DIN	Serial Data Input of the 16-bit shift register. Data is clocked into the register on the rising edge of SCLK.
2	CS̄	Chip Select (active low). Enables data to be shifted into the 16-bit shift register. Programming commands are executed at the rising edge of CS̄.
3	SCLK	Serial Clock Input. Data is clocked in on the rising edge of SCLK.
4	RESET	Asynchronous reset input (active low). Clears all registers to their default state (FFhex for DAC A and DAC B registers); all other registers are reset to 0 (including the input shift register).
5	V_{DD}	Positive Power Supply (2.7V to 5.5V). Bypass with 0.22µF to GND.
6	GND	Ground
7	V_{SS}	Negative Power Supply 0V or (-1.5V to -5.5V). Tie to GND for single supply operation. If a negative supply is applied, bypass with 0.22µF to GND.
8	OUTA	DAC A Output Voltage (Buffered). Resets to full scale. **Connect 0.05µF capacitor or greater to GND.**
9	OUTB	DAC B Output Voltage (Buffered). Resets to full scale. **Connect 0.01µF capacitor or greater to GND.**
10	OUTC	DAC C Output Voltage (Unbuffered). Resets to zero.
11	REFC	DAC C Reference Voltage
12	REFAB	DAC A/B Reference Voltage
13	I.C.	Internally connected. Do not make connections to this pin.
14	LOUT	Logic Output (latched)

FIGURE 6-2 MAX512/513 pin descriptions (*Maxim New Releases Data Book*, 1995, p. 9-14)

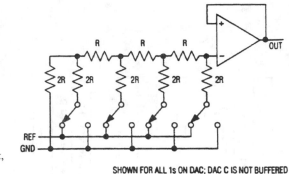

FIGURE 6-3
MAX512/513 simplified
circuit diagram (*Maxim
New Releases Data Book,*
1995, p. 9-15)

SHOWN FOR ALL 1s ON DAC; DAC C IS NOT BUFFERED

6.3 Reference Inputs versus DAC Output Range

The voltage at the reference inputs sets the full-scale output of the DACs in the usual manner. The input impedance of the reference inputs is code dependent. The lowest value, about 12 kohms for REF_C (8 kohms for REF_{AB}), occurs when the digital input code to be converted is 01010101 (which is 55 in hex code). The maximum input impedance (infinity) occurs when the input code is zero.

In the shutdown mode, the selected DAC output is set to zero. The value stored in the DAC register remains unchanged. This removes the load from the reference input to save power. Bringing the DACs out of shutdown restores the DAC output voltage. Because the input resistance at the reference is code dependent, the DAC reference sources should have an output impedance of no more than 5 ohms. The input capacitance at the reference input pins is also code dependent, and typically does not exceed 25 pF.

The reference voltage on REF_{AB} can be anywhere between the power-supply limits (or rails). In dual-supply mode, a positive input voltage on REF_{AB} should be less than (V_{DD} − 1.5 V) to avoid saturating the buffer amplifiers. The reference voltage includes the negative supply rail (see Section 6.4). The REF_C input accepts positive voltages up to V_{DD} and should not be forced below ground. The absolute difference between any reference voltage and GND should not exceed 6 V.

6.4 Output Buffer Amplifiers (DAC A/DAC B)

As shown in Fig. 6-1, the DAC A and DAC B voltage outputs are internally buffered. The buffer amplifiers have a rail-to-rail (V_{SS} to V_{DD}) output-voltage range. In single-supply mode, the DAC outputs A and B are internally divided by two and the buffer is set to a gain of two, eliminating the need for a buffer input-voltage range to the positive-supply rail. In dual-supply mode, the DAC outputs are not attenuated, and the buffer is set to unity gain.

Although only necessary for negative output voltages, the dual-supply mode can be used even if the desired DAC output voltage is positive. Possible errors associated with the divide-by-two attenuator and gain-of-two buffers in the single-supply mode are eliminated in dual-supply mode. However, do not use reference voltages higher than (V_{DD} − 1.5 V) for dual supply.

The DAC A output amplifier can source and sink up to 5 mA (0.5 mA for DAC B). The amplifier is unity-gain stable with a capacitive load of 0.05 μF (0.01 μF for DAC) or greater. The slew rate is limited by the load capacitor and is typically 0.1 V/μs with a 0.1-μF load (0.01 μF for DAC B). (If you are not familiar with amplifier terms such as *slew rate* and *unity gain*, read my *Simplified Design of IC Amplifiers* (Butterworth–Heinemann, 1996.)

6.5 Unbuffered Output (DAC C)

The output of DAC C is unbuffered and has a typical output impedance of 24 k. The output can be used to drive a high-impedance load, such as an op amp or comparator, and has 35-μs typical settling time to ½ LSB with a single 3-V supply. Use DAC C if a a quick, dynamic response is required.

6.6 Shutdown Mode

When programmed to shutdown mode, the outputs of DAC A and DAC B go into a high-impedance state where virtually no current flows into or out of the buffer amplifiers. The output of DAC C goes to 0 V when shut down. In shutdown, both reference inputs are high impedance (typically 2 M) to conserve current drain from the system reference. As a result, the system reference does not have to be powered down. The logic output (L_{OUT}) remains active in shutdown. When coming out of shutdown, the DAC outputs return to the values kept in the registers. The recovery time is equivalent to the DAC settling time.

6.7 Reset

The $\overline{RESET}$ input is active low. When asserted ($\overline{RESET}$ = 0), DAC A and DAC B are set to full scale (FF in hex) and active. DAC C is set to zero code (00 in hex) and active. The 16-bit serial register is cleared to 0000 in hex. L_{OUT} is reset to zero.

6.8 Serial Interface

Figures 6-4 and 6-5 show the basic serial-interface timing and input shift-register configuration, respectively. Figures 6-6 and 6-7 show the detailed timing and timing characteristics, respectively. An active-low chip select ($\overline{CS}$) enables the shift register to receive data bits from the serial data input. Data bits are clocked into the shift register on every rising edge of the serial-clock signal (SCLK). The clock frequency can be as high as 5 MHz.

Data bits are sent MSB first and can be transmitted in one 16-bit word. The write cycle can be interrupted at any time when $\overline{CS}$ is kept active (low). This allows two 8-bit-wide transfers, if desired. After all 16 bits are clocked into the input shift

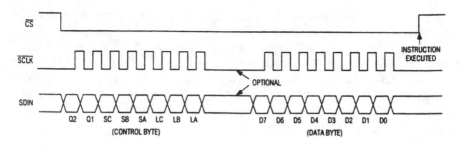

FIGURE 6-4 MAX512/513 basic serial-interface timing (*Maxim New Releases Data Book*, 1995, p. 9-16)

		B0*	DAC Data Bit 0 (LSB)
		B1	DAC Data Bit 1
		B2	DAC Data Bit 2
	DATA BITS	B3	DAC Data Bit 3
		B4	DAC Data Bit 4
		B5	DAC Data Bit 5
		B6	DAC Data Bit 6
		B7	DAC Data Bit 7 (MSB)
		LA	Load Reg DAC A, Active High
		LB	Load Reg DAC B, Active High
		LC	Load Reg DAC C, Active High
	CONTROL BITS	SA	Shut Down DAC A, Active High
		SB	Shut Down DAC B, Active High
		SC	Shut Down DAC C, Active High
		Q1	Logic Output
		Q2**	Uncommitted Bit

FIGURE 6-5
MAX512/513 input shift-register configuration (*Maxim New Releases Data Book*, 1995, p. 9-16)

* Clocked in last.
**Clocked in first.

register, the rising edge of $\overline{CS}$ updates the DAC outputs, the shutdown status, and the status of the logic output. Because of their single-buffered structure, DACs cannot be simultaneously updated to different digital values.

6.9 Data Format and Control Codes

Figure 6-8 describes the serial-input data format. The 16-bit input word consists of an 8-bit control byte and an 8-bit data byte. The 8-bit control byte is not decoded internally. Every control bit performs one function. Data bits are clocked in starting with Q2 (uncommitted bit), followed by the remaining control bits and the data byte. The LSB of the data byte (B0) is the last bit clocked into the shift register.

Figure 6-9 shows an example of a 16-bit input word. In this example, 128 in decimal (or 80 in hex) is loaded into the DAC registers A and B, which are both active.

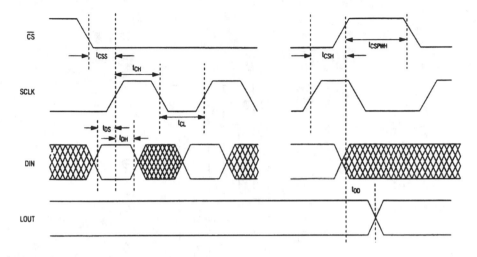

FIGURE 6-6 MAX512/513 detailed timing diagram (*Maxim New Releases Data Book*, 1995, p. 9-18)

DAC C is shutdown, and the contents of the DAC C register remain unchanged; LOUT is reset to 0.

6.10 Digital Inputs

The digital inputs are compatible with CMOS logic. Supply current increases slightly when the logic inputs are toggled through the transition zone between (0.3) (V_{DD}) and (0.7) (V_{DD}).

6.11 Digital Output

The latched digital output (L_{OUT}) has a 1.6-mA source capability with a (V_{DD} − 0.4 V) output level. With a 1.6-mA sink current, the output voltage at L_{OUT} is guaranteed to be no more than 0.4 V. The L_{OUT} signal can be used for digital auxiliary control. However, L_{OUT} remains fully active during shutdown mode.

6.12 Microprocessor Interfacing

The serial interface is compatible with Microwire™, SPI™, and QSPI™. For SPI™ and QSPI™, clear the CPOL and CPHA bits (CPOL = 0 and CPHA = 0). As described in SPI™ and QSPI™ literature, CPOL = 0 sets the inactive state of the clock to zero, and CPHA = 0 changes data at the falling edge of SCLK. This setting allows both SPI™ and QSPI™ to run at full clock speeds (0.5 MHz and 4 MHz, respectively).

TIMING CHARACTERISTICS (Note 4)

(V_{DD} = +4.5V to +5.5V for MAX512, V_{DD} = +2.7V to +3.6V for MAX513, V_{SS} = GND = 0V, T_A = T_{MIN} to T_{MAX}, unless otherwise noted.)

PARAMETER	SYMBOL	CONDITIONS	MIN	TYP	MAX	UNITS
SERIAL INTERFACE TIMING						
CS Fall to SCLK Rise Setup Time	tCSS		150			ns
SCLK Rise to CS Rise Setup Time	tCSH		150			ns
DIN to SCLK Rise Setup Time	tDS		50			ns
DIN to SCLK Rise Hold Time	tDH		50			ns
SCLK Pulse Width High	tCH		100			ns
SCLK Pulse Width Low	tCL		100			ns
Output Delay LOUT	tOD	C_L = 100pF			150	ns
CS Pulse Width High	tCSPWH		200			ns

Note 4: Guaranteed by design. Not production tested.

FIGURE 6-7 MAX512/513 timing characteristics (*Maxim New Releases Data Book*, 1995, p. 9-10)

CONTROL								DATA								FUNCTION
Q2	Q1	SC	SB	SA	LC	LB	LA	MSB B7	B6	B5	B4	B3	B2	B1	LSB B0	
•	•	•	•	•	0	0	0	X	X	X	X	X	X	X	X	No Operation to DAC Registers
•	•	•	•	•	1	0	0				8-Bit DAC Data					Load Register to DAC C
•	•	•	•	•	0	1	0				8-Bit DAC Data					Load Register to DAC B
•	•	•	•	•	0	0	1				8-Bit DAC Data					Load Register to DAC A
•	•	•	•	•	1	1	1				8-Bit DAC Data					Load All DAC Registers
•	•	0	0	0	•	•	•	X	X	X	X	X	X	X	X	All DACs Active
•	•	1	0	0	•	•	•	X	X	X	X	X	X	X	X	Shut Down DAC C
•	•	0	1	0	•	•	•	X	X	X	X	X	X	X	X	Shut Down DAC B
•	•	0	0	1	•	•	•	X	X	X	X	X	X	X	X	Shut Down DAC A
•	•	1	1	1	•	•	•	X	X	X	X	X	X	X	X	Shut Down All DACs
X	0	•	•	•	•	•	•	X	X	X	X	X	X	X	X	Reset LOUT
X	1	•	•	•	•	•	•	X	X	X	X	X	X	X	X	Set LOUT

X Don't care.
• Not shown for clarity. The functions of loading and shutting down the DACs and programming the logic can be combined in a single command.

FIGURE 6-8 MAX512/513 serial-interface data format (*Maxim New Releases Data Book*, 1995, p. 9-17)

Loaded in First														Loaded in Last	
Q2	Q1	SC	SB	SA	LC	LB	LA	B7	B6	B5	B4	B3	B2	B1	B0
X	0	1	0	0	0	1	1	1	0	0	0	0	0	0	0

The example above performs the following functions:
- 80hex (128 decimal) loaded into DAC registers A and B.
- Content of the DAC C register remains unchanged.
- DAC A and DAC B are active.
- DAC C is shut down.
- LOUT is reset to 0.

FIGURE 6-9 Example of 16-bit input word (*Maxim New Releases Data Book*, 1995, p. 9-17)

If the microprocessor does not have a serial port, three bits of a parallel port can be used to emulate a serial port by bit manipulation. Operate the serial clock only when necessary to minimize digital feedthrough at the voltage outputs.

6.13 Applications Data

The rest of this chapter describes how the MAX512/513 can be used for specific applications.

6.13.1 DAC with Bipolar Output

Figure 6-10 shows the MAX512/513 used with two Maxim ICL7612A op amps to provide a bipolar output. Figure 6-11 shows the bipolar code table. These op amps have rail-to-rail input common-mode range and rail-to-rail output-voltage swing, making the circuit of Fig. 6-10 ideal for a high output-voltage swing from low supply voltages.

There are two ways to obtain rail-to-rail outputs. First, operate the DACs with a single supply and a positive reference voltage. Second, use dual supplies with a positive or negative voltage at REF_{AB} and a positive voltage at REF_C. In either case, the op amps require dual supplies. When dual supplies are used, possible errors associated with the divide-by-two attenuator and gain-of-two buffer are eliminated (see Section 6.4).

With dual supplies, DAC A and DAC B can perform four-quadrant multiplication. In the dual-supply mode, REF_{AB} ranges from V_{SS} to ($V_{DD} - 1.5$ V). Because REF_C accepts only positive inputs, DAC C performs only two-quadrant multiplication.

6.13.2 Operating Voltages

Operating voltages for the circuit of Fig. 6-10 depend on the DAC used (MAX512 or MAX513). The MAX512 is fully specified to operation with $V_{DD} = 5$ V $\pm$ 10% and $V_{SS} = GND = 0$. The MAX513 is specified for single-supply

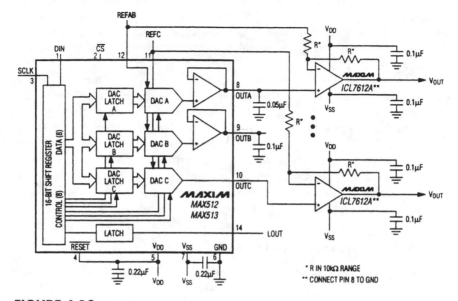

FIGURE 6-10 MAX512/513 used to provide a bipolar output (*Maxim New Releases Data Book*, 1995, p. 9-20)

DAC CONTENTS								ANALOG OUTPUT
B7	B6	B5	B4	B3	B2	B1	B0	
1	1	1	1	1	1	1	1	$+REF_- \times \left(\frac{127}{128}\right)$
1	0	0	0	0	0	0	1	$+REF_- \times \left(\frac{1}{128}\right)$
0	0	0	0	0	0	0	1	0V
1	1	1	1	1	1	1	0	$-REF_- \times \left(\frac{1}{128}\right)$
1	0	0	0	0	0	0	0	$-REF_- \times \left(\frac{127}{128}\right)$
0	0	0	0	0	0	0	0	$-REF_- \times \left(\frac{128}{128}\right) = -REF_-$

FIGURE 6-11

MAX512/513 bipolar code table (*Maxim New Releases Data Book*, 1995, p. 9-19)

Note :

$$1LSB = REF_- \times 2^{-(8-1)} = REF_- \times \left(\frac{1}{128}\right)$$

$$ANALOG\ OUTPUT = REF_- \times \left(\frac{D}{128} - 1\right)$$

operation with V_{DD} ranging from 2.7 V to 3.6 V, covering all commonly used supply voltages in 3-V systems. Both DACs can be used with a negative supply ranging from -1.5 V to 5.5 V. Using a negative supply typically improves zero-code error and settling time.

The two separate reference inputs for the DAC pair A/B and the unbuffered output C allow different full-scale output voltages. If a negative supply is used, the separate reference inputs allow different output polarities. In dual-supply mode, REF_{AB} can vary from V_{SS} to $(V_{DD} - 1.5$ V). In single-supply, the range for REF_{AB} is 0 V to V_{DD}. REF_C can range from GND to V_{DD}. Never force REF_C below ground.

Although power-supply sequencing is not critical, make sure that V_{SS} is never more than 0.3 V above ground if a negative supply is used. Also do not apply signals to the digital inputs until the DAC is powered up. If this is not possible, add current-limiting resistors to the digital inputs.

6.13.3 Bypasses and Grounds

In single-supply operation (V_{SS} = GND), GND and V_{SS} should be connected to the highest quality ground available. Bypass V_{DD} with a 0.1-µF to 0.22-µF capacitor to ground (see Fig. 6-10). For dual-supply operation, bypass V_{SS} with a 0.1-µF to 0.22-µF capacitor to GND. (Note that in Fig. 6-10, $\overline{RESET}$ is tied to V_{DD}, eliminating the need for an external reset signal.)

The reference inputs can be used without bypassing. However, for the best line-load transient response and noise performance, bypass the reference inputs with 0.1-µF to 4.7-µF capacitors to GND.

6.13.4 PC-Board Layout

As always, careful PC-board layout minimizes crosstalk among DAC output, reference inputs, and digital inputs. Separate analog lines by placing ground traces between the lines. Make sure that high-frequency digital lines are not routed in parallel to analog lines.

6.13.5 Unipolar Output

Figure 6-12 shows the code table when a unipolar output is required. With unipolar, the output voltage and the reference voltage have the same polarity. Both the MAX512 and MAX513 can be used with a single supply if the reference voltages are positive. With a negative supply, the REF_{AB} voltage can vary from V_{SS} to approximately $(V_{DD} - 1.5$ V), allowing two-quadrant multiplication.

6.13.6 RF Applications

Figure 6-13 shows how DAC A and DAC B can provide negative bias for two RF amplifiers using gallium-arsenide field-effect transistor, GaAs FETs. Such FETs require that the gate be negatively biased to ensure that there is no drain current. In a

DAC CONTENTS								ANALOG OUTPUT
B7	B6	B5	B4	B3	B2	B1	B0	
1	1	1	1	1	1	1	1	$+\text{REF}_- \times \left(\dfrac{255}{256}\right)$
1	0	0	0	0	0	0	1	$+\text{REF}_- \times \left(\dfrac{129}{256}\right)$
1	0	0	0	0	0	0	0	$+\text{REF}_- \times \left(\dfrac{128}{256}\right) = +\dfrac{\text{REF}_-}{2}$
0	1	1	1	1	1	1	1	$+\text{REF}_- \times \left(\dfrac{127}{256}\right)$
0	0	0	0	0	0	0	1	$+\text{REF}_- \times \left(\dfrac{1}{256}\right)$
0	0	0	0	0	0	0	0	0V

FIGURE 6-12

MAX512/513 unipolar code table (*Maxim New Releases Data Book*, 1995, p. 9-19)

Note :

$$1\text{LSB} = \text{REF}_- \times 2^{-8} = \text{REF}_- \times \left(\frac{1}{256}\right)$$

$$\text{ANALOG OUTPUT} = \text{REF}_- \times \left(\frac{D}{256}\right)$$

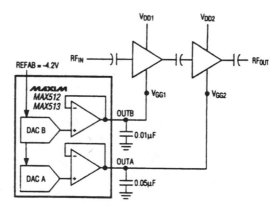

FIGURE 6-13

MAX512/513 used to provide negative bias for two GaAs FET RF amplifiers (*Maxim New Releases Data Book*, 1995, p. 9-20)

typical application, power to the RF amplifiers should not be turned on until the bias voltages provided by the DACs are fully established. The supply should be turned off before the bias voltage is switched off.

The DAC A and DAC B output can be used for controlling VCOs in mobile radios or cellular phones and to control for varactor and PIN-diode circuits. The unbuffered DAC C provides a span within GND and V_{DD} and is individually set at REF_C. A typical use for DAC C is to set or adjust offset and gain in an RF system.

CHAPTER 7

Simplified Design with Parallel-Input DAC

This chapter is devoted to simplified-design approaches for a typical DAC with parallel input. All the general design information in Chapters 1 through 3 applies to the examples in this chapter. The circuits in this chapter can be used immediately the way they are or, with alterations in component values, as a basis for simplified design of similar data-converter applications. The chapter concludes with a typical four-quadrant multiplication circuit.

7.1 General Description of DAC

Figure 7-1 shows the functional block diagram and pin configuration for the DAC (a MAX530). The IC is a low-power, 12-bit, voltage-output DAC that uses single +5-V or dual ±5-V supplies. The DAC has an on-chip voltage reference, and an output buffer-amplifier. Operating current is 250 mA from a single +5-V supply, making the DAC useful for portable and battery-powered applications. The available SSOP (shrink small outline package) measures only 0.1 square inch, using less board area than an 8-pin DIP. Twelve-bit resolution is achieved through laser trimming of the DAC, op amp, and references. No further adjustments are required.

Internal gain-setting resistors can be used to define a DAC output-voltage range of 0 V to +2.048 V, 0 V to +4.096 V, or ±2.048 V. Four-quadrant multiplication is possible without external resistors or op amps. The parallel logic inputs are double buffered and are compatible with 4-bit, 8-bit, and 16-bit microprocessors.

7.2 Data-Converter Operation

Figures 7-2 and 7-3 show the pin descriptions and simplified circuit diagram, respectively. The MAX530 consists of a parallel-input logic interface, a 12-bit R-2R ladder, a reference, and an op amp. Figure 7-1 shows the control lines and signal flow through the input data latch to the DAC latch and the 2.048-V reference and output op amp.

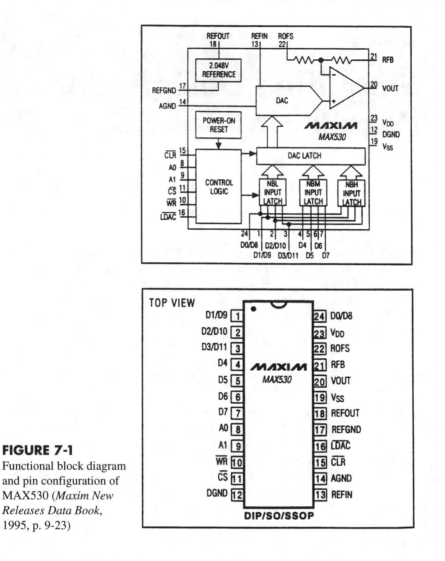

FIGURE 7-1

Functional block diagram and pin configuration of MAX530 (*Maxim New Releases Data Book*, 1995, p. 9-23)

7.3 R-2R Ladder

As shown in Fig. 7-3, the MAX530 uses an "inverted" R-2R ladder network with a BiCMOS op amp to convert 12-bit digital data into analog voltage levels. In a standard DAC (see Chapters 1 through 3), the REF_{IN} pin is the current output and is connected to the summing junction, or virtual ground, of an op amp. This makes the output voltage of a standard DAC the inverse of the reference voltage. In the inverted configuration of the MAX530, the ladder output voltage is of the same polarity as the reference input. This makes the MAX530 suitable for single-supply operation. The BiCMOS op amp is then used to buffer, invert, or amplify the ladder signal, as necessary.

PIN	NAME	FUNCTION
1	D1/D9	Input Data D1 if A0 = 0 and A1 = 1, or D9 if A0 = A1 = 1
2	D2/D10	Input Data D2 if A0 = 0 and A1 = 1, or D10 if A0 = A1 = 1
3	D3/D11	Input Data D3 if A0 = 0 and A1 = 1, or D11 (MSB) if A0 = A1 =1
4	D4	Input Data D4, or tie to D0 and multiplex if A0 = 1 and A1 = 0
5	D5	Input Data D5, or tie to D1 and multiplex if A0 = 1 and A1 = 0
6	D6	Input Data D6, or tie to D2 and multiplex if A0 = 1 and A1 = 0
7	D7	Input Data D7, or tie to D3 and multiplex if A0 = 1 and A1 = 0
8	A0	Address Line A0. With A1, used to multiplex 4 of 12 data lines to load low (NBL), middle (NBM), and high (NBH) 4-bit nibbles. (12 bits can also be loaded as 8+4.)
9	A1	Address Line A1. Set A0 = 1 and A1 = 1 for NBL, A0 = 1 and A1 = 0 for NBM, or A0 = A1 = 1 for NBH. See Table 2 for complete input latch addressing.
10	$\overline{\text{WR}}$	Write Input (active low). Used with $\overline{\text{CS}}$ to load data into the input latch selected by A0 and A1.
11	$\overline{\text{CS}}$	Chip Select (active low). Enables addressing and writing to this chip from common bus lines.
12	DGND	Digital Ground
13	REFIN	Reference Input. Input for the R-2R DAC. Connect an external reference to this pin or a jumper to REFOUT (pin 18) to use the internal 2.048V reference.
14	AGND	Analog Ground
15	$\overline{\text{CLR}}$	Clear (active low). A low on $\overline{\text{CLR}}$ resets the DAC latches to all 0s.
16	$\overline{\text{LDAC}}$	Load DAC Input (active low). Driving this asynchronous input low transfers the contents of the input latch to the DAC latch and updates VOUT.
17	REFGND	Reference Ground must be connected to AGND when using the internal reference. Connect to V_{DD} to disable the internal reference and save power.
18	REFOUT	Reference Output. Output of the internal 2.048V reference. Tie to REFIN to drive the R-2R DAC.
19	V_{SS}	Negative Power Supply. Usually ground for single-supply or -5V for dual-supply operation.
20	VOUT	Voltage Output. Op-amp buffered DAC output.
21	RFB	Feedback Pin. Op-amp feedback resistor. Always connect to VOUT.
22	ROFS	Offset Resistor Pin. Connect to VOUT for G = 1, to AGND for G = 2, or to REFIN for bipolar output.
23	V_{DD}	Positive Power Supply (+5V)
24	D0/D8	Input Data D0 (LSB) if A0 = 0 and A1 = 1, or D8 if A0 = A1= 1

FIGURE 7-2 MAX530 pin descriptions (*Maxim New Releases Data Book*, 1995, p. 9-29)

Ladder resistors are nominally 80 k to conserve power and are laser trimmed for gain and linearity. The input impedance at REF_{IN} is code dependent. When the DAC register is all 0s, all legs of the ladder are grounded and REF_{IN} is open or no-load. Maximum loading (minimum REF_{IN} impedance) occurs at code 010101 (or 555 in hex code). Minimum reference input impedance at this code is guaranteed to be not less than 40 k.

The REF_{IN} and REF_{OUT} pins allow the user to choose between driving the R-2R ladder with the on-chip reference or with an external reference. REF_{IN} can be below analog ground when dual supplies are used.

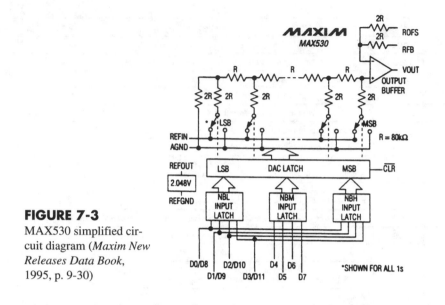

FIGURE 7-3

MAX530 simplified circuit diagram (*Maxim New Releases Data Book,* 1995, p. 9-30)

7.4 Internal Reference

The on-chip reference is laser trimmed to generate 2.048 V at REF$_{OUT}$. The output stage can source and sink current so REF$_{OUT}$ can settle to the correct voltage quickly in response to code-dependent loading changes. Typical source current is 5 mA, with a sink current of 100 µA.

REF$_{OUT}$ connects the internal reference to the R-2R DAC ladder at REF$_{IN}$. The R-2R ladder draws 50 µA maximum load current. If any other connection is made to REF$_{OUT}$, make certain that the total load current is less than 100 µA to avoid gain errors. A separate REF$_{GND}$ pin is provided to isolate reference currents from other analog and digital ground currents. When the internal reference is used, REF$_{GND}$ must be connected to AGND. In applications in which the internal reference is not used, connect REF$_{GND}$ to V$_{DD}$. This shuts down the reference and saves about 100 µA of V$_{DD}$ supply current.

7.5 Internal Reference Noise

Figure 7-4 shows the noise characteristics of the internal reference. To get the specified noise performance, connect a 33-µF capacitor from REF$_{OUT}$ to REF$_{GND}$. Using smaller capacitance values increases noise. Using values less than 3.3 µF can compromise stability of the reference. For lowest possible noise, insert a buffered RC filter between REF$_{OUT}$ and REF$_{IN}$.

7.6 Output Buffer-Amplifier

The output buffer-amplifier uses a folded-cascade input stage and a type AB output stage. (Those not familiar with amplifier circuits are invited to read *Simplified*

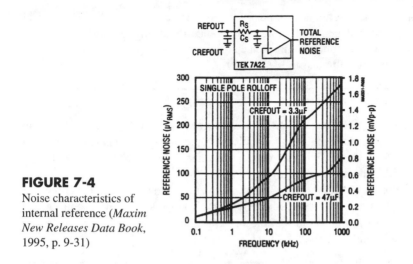

FIGURE 7-4

Noise characteristics of internal reference (*Maxim New Releases Data Book*, 1995, p. 9-31)

Design of IC Amplifiers, Butterworth–Heinemann, 1996.) Large-output devices with low series resistance allow the output to swing to ground in single-supply operation. The output buffer is unity-gain stable. Input-offset voltage and supply current are laser trimmed. Settling time is 25 µs to 0.01% of full scale. The output is short-circuit protected and can drive a 2-k load with more than 100 pF of load capacitance.

The output can be placed in unity-gain (G = 1), in a gain of two (G = 2), or in bipolar-output mode using the ROFS and RFB pins (Figs. 7-1 and 7-2). These pins are used to define the DAC output-voltage range of 0 V to +2.048 V, 0 V to +4.096 V, or ±2.048 V, by connecting ROFS to V_{OUT}, GND, or REF_{IN}, as shown in Fig. 7-5. RFB is always connected to V_{OUT}.

7.7 External Reference

If an external reference is required, the manufacturer recommends a MAX873A (2.5 V, ±15-mV initial accuracy, 7-ppm V°C maximum temperature coefficient). In any event, the external reference must be in the range (V_{SS} + 2 V) to (V_{DD} − 2 V) for dual-supply, unity-gain operation. If single-supply, unity-gain is used, the external reference must be positive and must not exceed (V_{DD} − 2 V).

FIGURE 7-5

ROFS connections (*Maxim New Releases Data Book*, 1995, p. 9-31)

ROFS CONNECTED TO:	DAC OUTPUT RANGE	OP-AMP GAIN
VOUT	0V to 2.048V	G = 1
AGND	0V to 4.096V	G = 2
REFIN	-2.048V to +2.048V	Bipolar

Note: Assumes RFB = VOUT and REFIN = REFOUT = 2.048V

As always, the reference voltage determines the DAC full-scale output. Because of the code-dependent nature of reference input impedances (see Section 7.3), a high-quality, low-output-impedance amplifier should be used to drive REF$_{IN}$. (The manufacturer recommends a MAX480.)

7.8 Reset Functions

An internal power-on reset (POR) circuit forces the DAC register to reset to all 0s when V$_{DD}$ is first applied. The POR pulse is typically 1.3 µs. However, it might take as long as 2 ms for the internal reference to charge the large filter capacitor and settle to the trimmed value.

In addition to POR, a clear ($\overline{CLR}$) pin, when held low, sets the DAC register to all 0s. $\overline{CLR}$ operates asynchronously and independently from chip select ($\overline{CS}$). With the DAC input at all 0s, the op-amp output is at 0 for unity-gain and G = 2 configurations.

7.9 Shutdown Mode

Figure 7-6 shows the MAX530 connected for low-current shutdown mode. The MAX530 is designed for low power consumption, and the circuit of Fig. 7-6 requires the minimum shutdown current because of the following. In single-supply mode (V$_{DD}$ = +5 V, V$_{SS}$ = GND), the initial supply current is typically 160 µA, including the reference, op amp, and DAC. This low current occurs when the power-on reset cir-

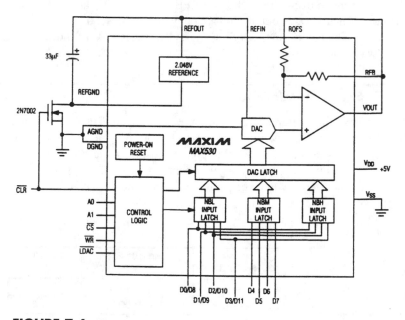

FIGURE 7-6 Connections for low-current shutdown mode (*Maxim New Releases Data Book*, 1995, p. 9-32)

cuit clears the DAC to all 0s and forces the op-amp output to zero. Under these conditions, there is no internal load on the reference (DAC = 000 in hex, REF$_{IN}$ = infinity) and the op amp operates at minimum quiescent current. The $\overline{CLR}$ signal resets the MAX530 to these same conditions and can be used to control a power-saving mode when the DAC is not being used by the system.

An additional 110 µA of supply current can be saved when the internal reference is not used by connecting REF$_{GND}$ to V$_{DD}$ (through $\overline{CLR}$), as shown in Fig. 7-6. The low on-resistance of the 2N7002 FET turns off the internal resistance. When $\overline{CLR}$ is high, the FET pulls REF$_{GND}$ to AGND, and both the reference and DAC operate normally. When $\overline{CLR}$ goes low, REF$_{GND}$ is pulled up to V$_{DD}$ and the reference is shut down. At the same time, $\overline{CLR}$ resets the DAC register to all 0s, and the op-amp output goes to 0 V for unity-gain and G = 2 operating modes. This reduces the total single-supply operating current from 250 µA (400 µA max) to typically 40 µA in the shutdown mode.

A small error voltage is added to the reference output by the reference current flowing through the FET. (The FET on-resistance should be less than 5 ohms.) A typical reference current of 100 µA adds about 0.5 mV to REF$_{OUT}$. Because the reference current and on-resistance increase with temperature, the overall temperature coefficient degrades slightly.

As data bits are loaded into the DAC and the output moves above GND, the op-amp quiescent current increases to the nominal value, and the total operating current averages 250 µA. Using dual supplies (±5 V), the op amp is fully biased continuously, and the V$_{DD}$ supply current is more constant at 250 µA. The V$_{SS}$ current is typically 150 µA.

The MAX530 logic inputs are compatible with TTL and CMOS. However, to get the lowest power dissipation, drive the digital inputs with rail-to-rail CMOS. With TTL logic levels, the power requirement increases by a factor of about 2.

7.10 Parallel Logic Interface

Figures 7-7 and 7-8 show the input addressing scheme and write-cycle timing diagram, respectively. The MAX530 uses 8 data pins and double-buffered logic inputs to load data as 4 + 4 + 4, or 8 + 4. This makes it possible to interface with

FIGURE 7-7

Input addressing scheme
(*Maxim New Releases
Data Book*, 1995, p. 9-32)

$\overline{CLR}$	$\overline{CS}$	$\overline{WR}$	$\overline{LDAC}$	A0	A1	DATA UPDATED
L	X	X	X	X	X	Reset DAC Latches
H	H	X	H	X	X	No Operation
H	X	H	H	X	X	No Operation
H	L	L	H	H	H	NBH (D8-D11)
H	L	L	H	H	L	NBM (D4-D7)
H	L	L	H	L	H	NBL (D0-D3)
H	H	H	L	X	X	Update DAC Only
H	L	L	X	L	L	NBL and NBM (D0-D7)
H	L	L	L	H	H	NBH and Update DAC

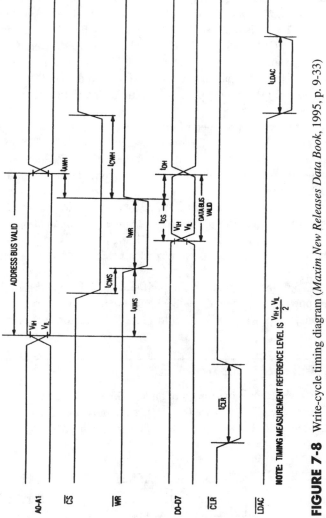

FIGURE 7-8 Write-cycle timing diagram (*Maxim New Releases Data Book*, 1995, p. 9-33)

4-bit, 8-bit, and 16-bit microprocessors. The 12-bit DAC latch is updated simultaneously through the control signal $\overline{\text{LDAC}}$. Signals A0, A1, $\overline{\text{WR}}$, and $\overline{\text{CS}}$ select which input latches to update.

The 12 data bits are broken down into nibbles (NB). NBL is the enable signal for the lowest 4 bits, NBM is the enable for the middle 4 bits, and NBH is the enable for the highest and most important 4 bits.

7.11 4-Bit Microprocessor Interface

Figures 7-9 and 7-10 show the basic connections and timing sequence, respectively, for a 4-bit microprocessor interface. The 4 low bits (D0-D3) are connected in parallel to the other 4 bits (D4-D7) and then to the microprocessor bus. Address lines A0 and A1 enable the input-data latches for high, middle, or low data nibbles. The

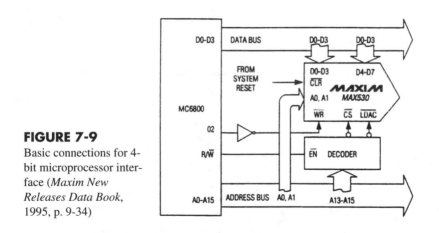

FIGURE 7-9
Basic connections for 4-bit microprocessor interface (*Maxim New Releases Data Book*, 1995, p. 9-34)

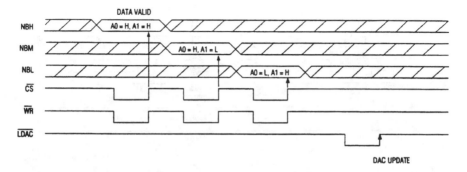

FIGURE 7-10 Timing sequence for 4-bit microprocessor interface (*Maxim New Releases Data Book*, 1995, p. 9-34)

microprocessor sends $\overline{\text{CS}}$ and $\overline{\text{WR}}$) signals to latch in each of three nibbles in three cycles when the data bits are valid.

7.12 8-Bit or 16-Bit Microprocessor Interface

Figures 7-11 and 7-12 show the basic connections and timing sequence, respectively, for an 8-bit or 16-bit microprocessor interface, using the $\overline{\text{LDAC}}$ function. Figure 7-13 shows the timing sequence when the $\overline{\text{LDAC}}$ function is not used (held low or 0).

With $\overline{\text{LDAC}}$ held high, the user can load NBH, or NBL + NBM, in any order, as shown in Fig. 7-12. For the fastest throughput, use the sequence in Fig. 7-13, in which $\overline{\text{LDAC}}$ is held low. Address lines A0 and A1 are tied together and the DAC is loaded in 2 cycles as 8 + 4. (The DAC latch is transparent in this mode.) Always load NBL and NBM first, followed by NBH.

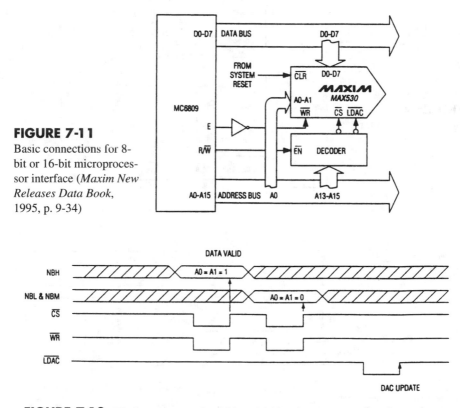

FIGURE 7-11

Basic connections for 8-bit or 16-bit microprocessor interface (*Maxim New Releases Data Book*, 1995, p. 9-34)

FIGURE 7-12 Timing sequence for 8-bit or 16-bit microprocessor interface using $\overline{\text{LDAC}}$ function (*Maxim New Releases Data Book*, 1995, p. 9-34)

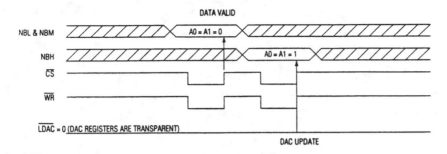

FIGURE 7-13 Timing sequence for 8-bit or 16-bit microprocessor interface with $\overline{\text{LDAC}} = 0$ (*Maxim New Releases Data Book*, 1995, p. 9-35)

$\overline{\text{LDAC}}$ is asynchronous with respect to $\overline{\text{WR}}$. If $\overline{\text{LDAC}}$ is brought low before or at the same time $\overline{\text{WR}}$ goes high, $\overline{\text{LDAC}}$ must remain low for at least 50 ns to ensure that the correct data bits are latched. Data bits are latched into the DAC registers on the $\overline{\text{LDAC}}$ rising edge.

7.13 Unipolar Operation

Figures 7-14 and 7-15 show the basic connections for unipolar operation. Figures 7-16 and 7-17 show the corresponding unipolar code tables.

As shown in Figs. 7-5 and 7-14, the MAX530 is configured for a 0-V to +2.048-V unipolar output when ROFS and RFB are connected to V_{OUT}. The IC can operate from either single or dual supplies in this configuration. Figure 7-16 shows the DAC-latch contents (input) versus the analog V_{OUT} (output), where 1 LSB = $REF_{IN} (2^{-12})$.

As shown in Figs. 7-5 and 7-15, the MAX530 is configured for a 0-V to +4.096-V unipolar output when ROFS is connected to AGND and RFB is connected

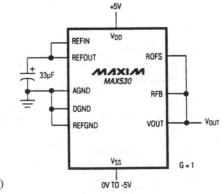

FIGURE 7-14

Basic connections for unipolar operation with 0-V to +2.048-V output (*Maxim New Releases Data Book*, 1995, p. 9-35)

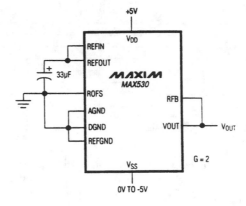

FIGURE 7-15

Basic connections for
unipolar operation with
0-V to +4.096-V output
(*Maxim New Releases
Data Book*, 1995, p. 9-35)

INPUT			OUTPUT
1111	1111	1111	$(VREF) \dfrac{4095}{4096}$
1000	0000	0001	$(VREF) \dfrac{2049}{4096}$
1000	0000	0000	$(VREF) \dfrac{2048}{4096} = +VREF/2$
0111	1111	1111	$(VREF) \dfrac{2047}{4096}$
0000	0000	0001	$(VREF) \dfrac{1}{4096}$
0000	0000	0000	0V

FIGURE 7-16

Unipolar code table for
0-V to +2.048-V output
(*Maxim New Releases
Data Book*, 1995, p. 9-36)

INPUT			OUTPUT
1111	1111	1111	$+2 \, (VREF) \dfrac{4095}{4096}$
1000	0000	0001	$+2 \, (VREF) \dfrac{2049}{4096}$
1000	0000	0000	$+2 \, (VREF) \dfrac{2048}{4096} = +VREF$
0111	1111	1111	$+2 \, (VREF) \dfrac{2047}{4096}$
0000	0000	0001	$+2 \, (VREF) \dfrac{1}{4096}$
0000	0000	0000	0V

FIGURE 7-17

Unipolar code table for
0-V to +4.096-V output
(*Maxim New Releases
Data Book*, 1995, p. 9-36)

to V_{OUT}. The IC can operate from either single or dual supplies in this configuration. Figure 7-17 shows the DC-latch contents versus analog V_{OUT}, where 1 LSB = $(2)(REF_{IN})(2^{-12}) = (REF_{IN})(2^{-11})$.

7.14 Bipolar Operation

Figures 7-18 and 7-19 show the basic connections and code table, respectively, for bipolar operation. As shown in Figs. 7-5 and 7-18, the MAX530 is configured for -2.048-V to $+2.048$-V bipolar range when ROFS is connected to REF_{IN} and RFB is connected to V_{OUT}. Dual supplies (± 5 V) must be used in this configuration. Figure 7-19 shows the DAC-latch contents versus analog V_{OUT}, where 1 LSB = $REF_{IN}(2)^{-11}$).

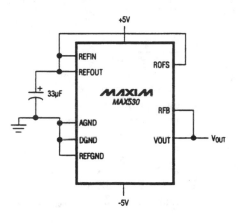

FIGURE 7-18

Basic connections for bipolar operation (*Maxim New Releases Data Book*, 1995, p. 9-37)

INPUT			OUTPUT
1111	1111	1111	$(+VREF)\frac{2047}{2048}$
1000	0000	0001	$(+VREF)\frac{1}{2048}$
1000	0000	0000	0V
0111	1111	1111	$(-VREF)\frac{1}{2048}$
0000	0000	0001	$(-VREF)\frac{2047}{2048}$
0000	0000	0000	$(-VREF)\frac{2048}{2048} = -VREF$

FIGURE 7-19

Bipolar code table (*Maxim New Releases Data Book*, 1995, p. 9-36)

7.15 Four-Quadrant Multiplication

Figure 7-20 shows the basic connections for four-quadrant multiplication. This configuration requires that ROFS be connected to REF$_{IN}$, that an offset-dual or two's-complement code be used, bipolar power supplies be available, and the bipolar analog input at REF$_{IN}$ (for multiplication) be in the range of V_{SS} + 2 V to V_{DD} − 2V.

A 12-bit DAC output is typically (D)(VREF$_{IN}$)(G) where G is the gain (1 or 2) and D is the binary representation of the digital input divided by 2^{12} or 4096. This formula is precise for unipolar operation. However, for bipolar, two's-complement operation, the MSB is really a polarity bit. No resolution is lost because there are the same number of steps. However, the output voltage has been shifted. For example, 0V to 4.096 V (G = 2) is shifted to a range of −2.048 V to +2.048 V. Keep in mind that when the DAC is used as a four-quadrant multiplier, the scale is skewed. The negative full scale is −VREF$_{IN}$, with the positive full scale at +VREF$_{IN}$ − 1 LSB.

7.16 Single-Supply Problems

As in the case of any op amp, there are linearity problems when a single supply is used, and these problems apply to the output op amp of the MAX530. For example, op amp output offset can be positive or negative. There is no major problem when the output offset is positive because this can be corrected (usually by an offset adjustment at the input).

The offset problem becomes aggravated when the offset is negative, and there is only one supply. The output cannot follow in a linear manner when there is no negative supply. In that case, the amplifier output (V$_{OUT}$) remains at ground until the DAC

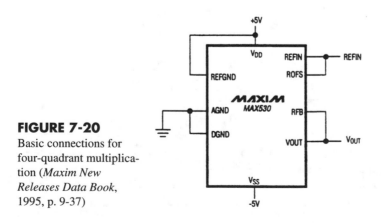

FIGURE 7-20

Basic connections for four-quadrant multiplication (*Maxim New Releases Data Book*, 1995, p. 9-37)

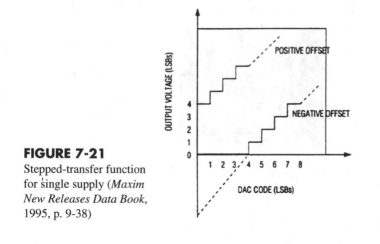

FIGURE 7-21

Stepped-transfer function for single supply (*Maxim New Releases Data Book*, 1995, p. 9-38)

voltage is sufficient to overcome the offset and the offset becomes positive. This results in a stepped transfer function such as shown in Fig. 7-21.

Linearity normally is measured after one allows for zero error and gain error. However, because the actual value of a negative offset is unknown, the value cannot be accounted for during testing in single-supply operation. In the MAX530, linearity and gain error are measured from code 11 to code 4095. The output-amplifier offset does not affect monotoniticy, and the DACs are guaranteed monotonic starting with code zero. In dual-supply operation, linearity and gain are measured from code 0 to 4095.

7.17 Bypasses, Grounds, and PC-Board Layout

All of the recommendations in Sections 6.13.3 and 6.13.4 apply to the MAX530 (and to most DACs).

7.18 Digital and Analog Feedthrough

High-speed data at any of the digital-input pins can couple through a DAC and cause internal stray capacitance to appear as noise at the DAC output, even though $\overline{LDAC}$ and $\overline{CS}$ are held high. (This is known as digital feedthrough.) The condition is tested by holding $\overline{LDAC}$ and $\overline{CS}$ high and toggling the data inputs from all 1s to all 0s. A typical pattern is shown in Fig. 7-22.

Because of internal stray capacitance, higher-frequency analog input signals at REF_{IN} (for example, for four-quadrant multiplication) can couple to the output, even when the input digital code is all 0s. (This is known as analog feedthrough.) The condition is tested by sweeping REF_{IN} and setting CLR to low (which sets the DAC switches to 000 in hex). Typical test results are shown in Fig. 7-23.

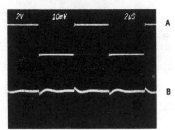

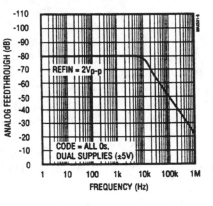

FIGURE 7-22
Typical digital-
feedthrough pattern
(*Maxim New Releases
Data Book*, 1995, p. 9-28)

FIGURE 7-23
Analog feedthrough ver-
sus frequency (*Maxim
New Releases Data Book*,
1995, p. 9-27)

CHAPTER **8**

Simplified Design Approaches

This chapter is devoted to simplified design approaches for a cross section of data-converter ICs. All of the general design information in Chapters 1 through 3, as well as the specific design information of Chapters 4 through 7, applies to the examples in this chapter. However, each IC has special design requirements, all of which are discussed in detail. The circuits in this chapter represent both classic and state-of-the-art applications. In addition, the circuits can be used immediately the way they are or, with alterations in component values, as a basis for simplified design of similar data converters. If the circuits do not perform as expected (unthinkable), consult the testing and troubleshooting information at the end of Chapter 1.

8.1 References for ADCs and DACs

Although many of the data-converter ICs described in this book have a built-in reference, some require external references (or will work better with a precision external reference). This section is devoted to simplified design of precision, low-drift references.

Even if conversion linearity is perfect, the accuracy of any converter is limited by the temperature drift or long-term drift of the voltage reference. If the voltage reference is allowed to add ½ LSB to the converter, a reference must be reasonably stable even for small temperature variations. When temperature changes are large, the reference accuracy becomes critical. Figure 8-1 shows a comparison of reference requirements for various bit lengths in both data converters and digital panel meters.

A voltage reference used in a data converter must do several things besides supplying a fixed voltage. First, input power supply changes must be rejected by the reference. The zener used in the reference must be biased properly. (Other parts of the reference circuit scale the zener voltage and provide a low-impedance output.) The reference must reject ambient temperature changes so that the temperature drift of the reference, plus the zener drift, does not exceed the desired drift limit.

Although zener TC (temperature coefficient) is critical to reference performance, other sources of drift can easily add as much error as the zener, even in references with a modest 20 ppm/°C temperature drift or TC. Zener drift and op-amp

TABLE I. Maximum Allowable Reference Drift for 1/2 Least Significant Bits Error of Binary Coded Converter

TEMP CHANGE	BITS					
	6	8	10	12	14	
25°C	310	80	20	5	1.25	ppm/°C
50°C	160	40	10	2.5	0.6	ppm/°C
100°C	80	20	5	1.2	0.3	ppm/°C
125°C	63	16	3	1	0.2	ppm/°C

TABLE II. Maximum Allowable Reference Drift for 1/2 Digit Error of Digital Meters

TEMP CHANGE	DIGITS								
	2	2½	3	3½	4	4½	5	5½	
25°C	200	100	20	10	2	1	0.2	0.1	ppm/°C
5°C			100	50	10	5	1	0.5	ppm/°C

*0.01%/°C = 100 ppm/°C, 0.001%/°C = 10 ppm/°C, 0.0001%/°C = 1 ppm/°C

FIGURE 8-1 Comparison of reference requirements for various bit lengths (National Semiconductor, *Linear Applications Handbook,* 1994, p. 378)

DEVICE	ERROR	10V OUTPUT DRIFT
Zener	**Zener Drift**	
LM199A	0.5 ppm/°C	0.5 ppm/°C
LM199, LM399A	1 ppm/°C	1 ppm/°C
LM399	2 ppm/°C	2 ppm/°C
1N829, LM3999	5 ppm/°C	5 ppm/°C
LM129, 1N823A, 1N827A, LM329A	10–50 ppm/°C	10–50 ppm/°C
LM329, 1N821, 1N825	20–100 ppm/°C	20–100 ppm°C
Op Amp	**Offset Voltage Drift**	
LM725, LH0044, LM121	1 μV/°C	0.15 ppm/°C
LM108A, LM208A, LM308A	5 μV/°C	0.7 ppm/°C
LM741, LM101A	15 μV/°C	2 ppm/°C
LM741C, LM301A, LM308	30 μV/°C	4 ppm/°C
Resistors	**Resistance Ratio Drift**	
1% (RN55D)	50–100 ppm	20–40 ppm/°C
0.1% (Wirewound)	5–10	2–4 ppm
Tracking 1 ppm Film or Wirewound	—	0.4 ppm/°C

FIGURE 8-2 Drift-error contribution from various components (National Semiconductor, *Linear Applications Handbook,* 1994, p. 379)

drift add directly to the drift error. The error of resistors used in the reference circuit is a function of how well the scaling resistors track. (Resistors with a high TC can be used in precision references, if the scaling resistors track each other.)

Figure 8-2 shows the drift-error contribution from various components used in a 10-V reference (with a 6.9-V zener). The drift contribution of resistor mistracking is about 0.4, because the gain is 1.4. Figure 8-2 does not show the contribution of input supply variations. As a guideline, such variations can be ignored if the input is 1% regulated and the resistor feeding the zener is stable to 1%.

8.1.1 Voltage Reference with 20-ppm/°C Performance

Figures 8-3, 8-4, and 8-5 show three adjustable voltage references, all with a 20-ppm/°C (or better) temperature drift. The following is a summary of the design considerations.

In the circuit of Fig. 8-3, an LM308 op amp is used to increase the zener output to 10 V. This combination adds a worst-case drift of 4 ppm/°C to the 10 ppm/°C drift of the zener. Resistors R3 and R4 must track to better than 10 ppm/°C, bringing the total error to about 18 ppm/°C.

Potentiometer R5 and resistor R2 are included so that the output can be adjusted to eliminate the initial zener tolerance. The loading on R5 by R2 is small, and there is no tracking requirement between R5 and R2. However, it is necessary for R2 to track R3/R4 within 50 ppm.

In the circuit of Fig. 8-4, a low-drift reference and op amp are used to give a total drift (exclusive of resistors) of 3 ppm/°C. This relaxes the resistor-tracking requirement to about 50 ppm/°C, allowing ordinary 1% resistors to be used. If the circuit is used in applications requiring 3-ppm/°C to 5-ppm/°C overall drift, tighten the resistor-tracking. For more accurate applications, use Kelvin sensing for both output and ground.

For the lowest possible drift in either circuit (Fig. 8-3 or 8-4), substitute a 1-μV/°C op amp, 1-ppm tracking resistors, and an LM199A, using the components shown in Fig. 8-2. Such combinations can produce overall drifts of 1 ppm/°C. In both circuits, it is important to remember that the tracking of resistors can, at worst case, be twice the temperature drift of either resistance.

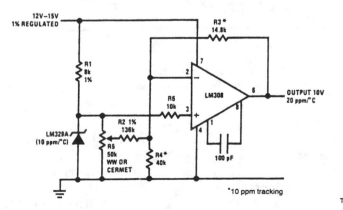

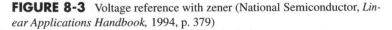

FIGURE 8-3 Voltage reference with zener (National Semiconductor, *Linear Applications Handbook,* 1994, p. 379)

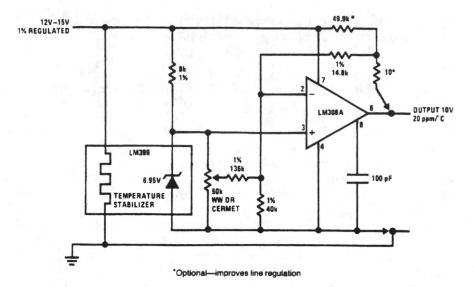

FIGURE 8-4 Voltage reference with temperature stabilizer (National Semiconductor, *Linear Applications Handbook,* 1994, p. 380)

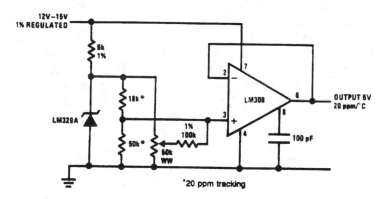

FIGURE 8-5 Voltage reference below zener value (National Semiconductor, *Linear Applications Handbook,* 1994, p. 380)

The circuit of Fig. 8-5 is used when the reference output is less than the zener voltage. In this case, the reference produces a 5-V output with a 12- to 15-V regulated (1%) input. The zener drift contributes proportionally to the output drift, whereas op-amp-offset drift adds a greater rate. With the 10-V references (Figs. 8-3 and 8-4), 15-μV/°C from the op amp contributes 2 ppm/°C. In the 5-V reference of Fig. 8-5, 15-μV/°C adds 3 ppm/°C. This makes the op-amp choice more important when the output voltage is lowered. Of course, if a high output impedance to the data converter can be tolerated (usually not), the op amp can be eliminated.

8.1.2 Voltage Reference with 1-ppm/°C Performance

Figure 8-6 shows a circuit that can be trimmed to provide 1-ppm/°C (or better) performance. The trimming procedure is as follows. Note that both the 40-k and 14.8-k resistors must be 1 ppm tracking.

Disconnect the zener and ground the op-amp input. Null the op-amp offset to zero using the 20-k wire-wound pot. Reconnect the zener, and adjust the circuit output to precisely 10 V (at pin 6 of LM108A) using the 50-k pot. Make a run through the full temperature range and note the drift. The LM121 will drift 3.8-μV/°C for every 1 μV of offset. So for every 5-μV/°C drift at the output, adjust the op amp 1 μV (1.4 μV measured at the circuit output) in the opposite direction with the 20-k pot. Readjust the circuit output to 10 V (50-k pot), and check drift through the full temperature range.

To get the best results, cycle the circuit through the temperature range a few times before making the final test or trimming. This relieves stress on components. Oven testing can sometimes cause thermal gradients across circuits, resulting in 50-μV to 100-μV errors. However, with careful layout and trimming, overall reference drifts of 0.1 to 0.2 ppm/°C are possible.

As always, good single-point grounding is important. Traces on a PC board can easily have 0.1-ohm resistance. Only 10 mA will cause a 1-μV shift. In addition, because these references are close to high-speed digital circuits, shielding might be necessary to prevent pickup at the op-amp inputs. Transient response to pickup, or

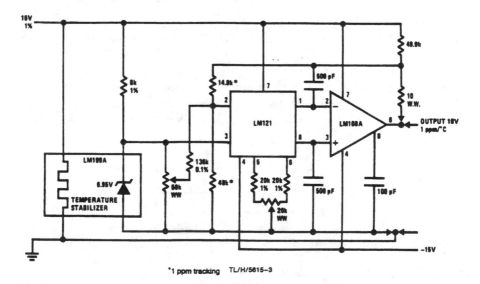

FIGURE 8-6 Voltage reference with 1-ppm/°C performance (National Semiconductor, *Linear Applications Handbook,* 1994, p. 381)

rapid-loading changes, can sometimes be improved with a large capacitor (1-μf to 10-μF) directly on the op-amp output. Of course, this depends on the op-amp stability.

8.2 Unusual ADC Applications

Figure 8-7 shows an ADC capable of providing solutions to many system-design problems. The following paragraphs summarize typical applications for the ADC.

8.2.1 Accommodating Arbitrary Zero and Span

In many systems, the analog signal to be converted does not range fully to ground (0.00 VDC), nor does the signal reach up to the full supply or reference-voltage value. This presents two problems for ADCs. First, a zero-offset is needed, and this offset might be in volts rather than the usual few millivolts provided. Second, the full-scale must be adjusted to accommodate this reduced span. (Span is the actual range of the analog input signal from V_{IN} MIN to V_{IN} MAX.)

When V_{IN} equals V_{IN} MIN, the differential input to the ADC is 0V, and a digital-output code of 0 is produced. When V_{IN} equals V_{IN} MAX, the differential input to the ADC is equal to the span. In the ADC of Fig. 8-7, there is an internal gain of two for the voltage, which is applied to pin 9 (the $V_{REF}/2$ input). This makes it possible to adjust the ADC to provide digital full scale and thus accommodate a wide range of analog-input voltages.

8.2.2 Analog Inputs Less Than Complete Supply Range

Figure 8-8 shows the ADC operating with a ratiometric transducer, which outputs 15% to 85% of the full 5-V supply (VCC). The transducer output is applied to V_{IN} (+), with 15% of VCC (0.75 V) applied to V_{IN} (−). The $V_{REF}/2$ pin is biased at one-half of the span, or ½ 85% − 15%), or 35% of the supply (1.75 V) through a unity-gain op amp. This circuit configuration can provide 9-bit performance with an 8-bit converter, if the span of the analog input uses only one-half of the available 0-V to 5-V span. (This would be a span of about 2.5 V, which could start anywhere over the range of 0 V to 2.5 V.)

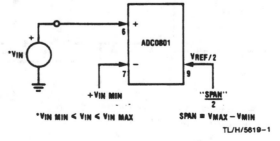

FIGURE 8-7 Basic circuit for arbitrary zero and span (National Semiconductor, *Linear Applications Handbook,* 1994, p. 465)

Note that linearity error increases when the analog-input span is reduced. This is shown in Fig. 8-9, which gives a comparison of three full-scale values (A = 5 V, B = 1.25 V, and C = 0.3125 V). Of course, resolution increases when the span is reduced. For example, when full scale is 5 V, 1 LSB is 20 mV, and the resolution is 8 bit. With full scale at 1.25 V, 1 LSB = 5 mV, and resolution is 10 bit. When full scale is reduced to 0.3125 V, 1 LSB is 1.22 mV, and resolution is 12 bit.

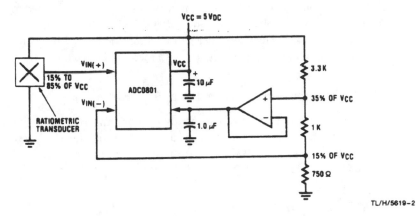

TL/H/5619-2

FIGURE 8-8 Operating with a ratiometric transducer (National Semiconductor, *Linear Applications Handbook,* 1994, p. 465)

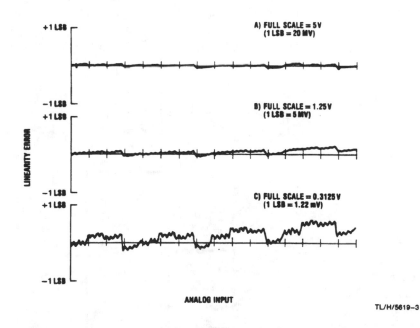

TL/H/5619-3

FIGURE 8-9 Linearity error for reduced full-scale spans (National Semiconductor, *Linear Applications Handbook,* 1994, p. 466)

8.2.3 10-Bit Resolution from 8 Bits

Figure 8-10 shows the basic connections to produce 10-bit resolution from the 8-bit ADC. The two extra bits are provided by the 2-bit external DAC (the resistor string) and the analog switch SW1. With these connections the $V_{REF}/2$ pin is supplied with $\frac{1}{8} V_{REF}$ so that each of the four spans encoded will be $2 \times \frac{1}{8} V_{REF} = \frac{1}{4} V_{REF}$.

In a practical circuit, the switch is replaced by an analog multiplexer (such as a classic CD4066 quad bilateral switch), and a microprocessor is programmed to do a binary search for the two MS bits. These two bits plus the 8 LSBs provided by the ADC result in the 10-bit data. This basic idea can be simplified to a 1-bit ladder to cover a particular range of analog input voltages with increased resolution.

8.2.4 Direct Encoding of Low-Level Signals

Figure 8-11 shows the ADC0801 connected to encode low-level signals without an external op amp. The V_{IN} (−) input is used as an offset adjustment. When the analog input applied at V_{IN} (+) is at 0 V, the 8-bit digital output is at all 0s. All 1s are output by the ADC when the analog input is at 500 mV.

8.2.5 Digitizing a Current Flow

Figure 8-12 shows the ADC0801 connected to provide a digital output that corresponds to a load current. This is done by sampling the load current with a 0.1-ohm resistor connected between VCC and the load. The voltage produced across the resistor is monitored at the V_{IN} (−) input. The V_{IN} (+) input is used as the offset adjustment. When the load current is 0, the digital output is at all 0s. All 1s are output by

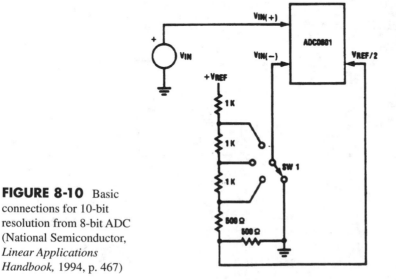

FIGURE 8-10 Basic connections for 10-bit resolution from 8-bit ADC (National Semiconductor, *Linear Applications Handbook,* 1994, p. 467)

TL/H/5619–4

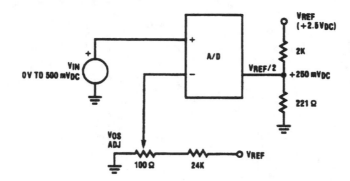

FIGURE 8-11 Direct encoding of low-level signals (National Semiconductor, *Linear Applications Handbook,* 1994, p. 468)

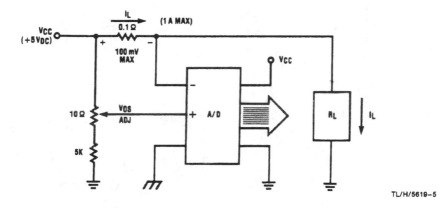

TL/H/5619–5

FIGURE 8-12 Digitizing a current flow (National Semiconductor, *Linear Applications Handbook,* 1994, p. 468)

the ADC when the load current is 1 A. (Do not exceed 100 mV across the load-sampling resistor.)

8.3 Data Acquisition with ADCs

Figure 8-13 shows ADCs specifically designed for data-acquisition applications. Figures 8-14 and 8-15 show the analog input-selection code and timing diagram, respectively. The ADC0816 and ADC0817, CMOS 16-channel data-acquisition devices are actually selectable multi-input 8-bit ADCs. However, in addition to a standard 8-bit SAR-type ADC, these devices also contain a 16-channel analog multiplexer (MUX) with 4-bit latched address inputs. As a result, the ICs include much of the circuitry required to build an 8-bit accurate, medium-throughput data-acquisition system.

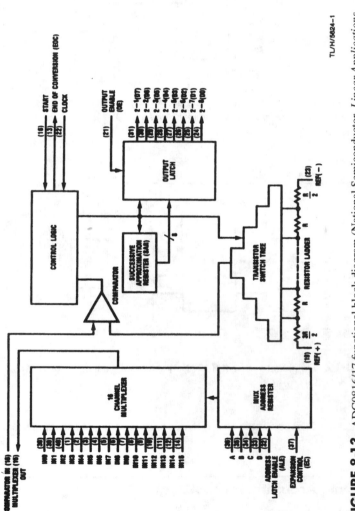

FIGURE 8-13 ADC0816/17 functional block diagram (National Semiconductor, *Linear Applications Handbook*, 1994, p. 590)

Address				Expansion	Selected
D	C	B	A	Control	Channel
0	0	0	0	1	IN0
0	0	0	1	1	IN1
0	0	1	0	1	IN2
0	0	1	1	1	IN3
0	1	0	0	1	IN4
0	1	0	1	1	IN5
0	1	1	0	1	IN6
0	1	1	1	1	IN7
1	0	0	0	1	IN8
1	0	0	1	1	IN9
1	0	1	0	1	IN10
1	0	1	1	1	IN11
1	1	0	0	1	IN12
1	1	0	1	1	IN13
1	1	1	0	1	IN14
1	1	1	1	1	IN15
X	X	X	X	0	NONE

FIGURE 8-14 Analog input-selection code (National Semiconductor, *Linear Applications Handbook,* 1994, p. 591)

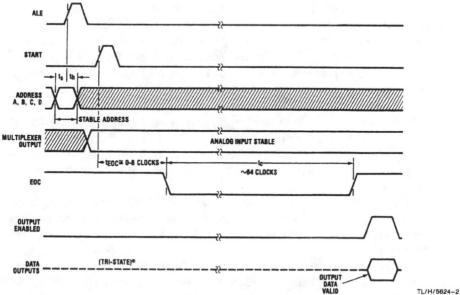

FIGURE 8-15 Timing diagram (National Semiconductor, *Linear Applications Handbook,* 1994, p. 591)

Although similar to other ADCs used in data-acquisition systems, these ICs have externally available multiplexer output and ADC-comparator input. This feature is useful when connecting signal-processing circuits to the ADC. Also, these ICs have an expansion-control pin to allow addition of more multiplexers, producing more input channels. Although the following information applies to specific ADCs, most of the principles involved can be applied to any ADC used in data-acquisition applications.

The ADC0816 is identical to the ADC0817 except for accuracy. The ADC0816 is the more accurate device, having a total unadjusted error of $\pm\frac{1}{2}$ LSB. The ADC0817 has a total unadjusted error of ± 1 LSB (and is, as one might expect, less expensive).

8.3.1 Functional Description

As shown in Fig. 8-13, the ADCs can be divided into two major functional blocks: a multiplexer-latch and an ADC. The multiplexer-latch is composed of a 16-channel multiplexer, a 4-bit channel-select register, and some channel-select decoding circuits. The channel-select address (Fig. 8-14) is loaded on the positive transition of the address latch enable (ALE) input (Fig. 8-15). A multiplexer-enable pin called expansion control (EC) is also provided. Taking the EC pin low disables the on-chip multiplexer, allowing other multiplexers to be used, expanding the number of inputs.

The output of the multiplexer usually feeds the ADC input. The ADC is composed of the usual comparator, 256R-type resistor ladder, SAR, control logic, and output data latch. During normal operation, the ADC control logic first detects a positive-going pulse on the START input. On the rising edge of this pulse, the internal registers are cleared, and they remain clear as long as START is high. When START goes low, the conversion is initiated. The control logic cycles to the beginning of the next approximation cycle, at which time EOC (end of conversion) goes low and the actual conversion is started.

During a conversion, the control logic selects a tap on the resistor ladder, and routes the signal through a transistor switch tree to the input of the comparator. The comparator then decides whether the tap voltage is higher or lower than the input signal and indicates the decision to the control logic. The control logic then decides which tap is to be selected next.

During this process, the SAR maintains a record of the conversion sequence. As shown in Fig. 8-15, it takes 8 clock periods per approximation and requires eight approximations to convert 8 bits. Thus 64 clock periods are required for a complete conversion.

After the entire conversion is complete, the data bits in the SAR are loaded into the output register. This register requires that the outputs be enabled by the OE (output enabled) input. The data bits can then be read by the control logic.

During operation, the EOC output must be monitored to determine whether the device is actively converting or is ready to output data. When the channel address is loaded, a positive-going pulse on START starts the conversion and causes EOC to fall. When EOC goes high again, the data bits are ready to be read (by raising the OE input). The data bits can be read any time prior to one clock period before completion of the next conversion.

8.3.2 Ratiometric Conversion

Figure 8-16 shows the ADCs connected for ratiometric conversion. Because both ends of the 256R resistor ladder are available externally, the ADCs are ideally suited for use with ratiometric transducers. As discussed in Chapters 1 through 3, a ratiometric transducer is a conversion device in which the output is proportional to some arbitrary full-scale value. The actual value of the transducer output is of no great importance, but the ratio of this output to the full-scale reference is valuable. The prime advantage of a ratiometric transducer is that an accurate reference is not essential. However, the reference should be noise free, because voltage spikes during a conversion could cause inaccurate results.

The circuit of Fig. 8-16 uses the existing 5-V supply for the reference, thus eliminating the need for a special external reference. Care should be taken to reduce power-supply noise. The supply lines should be well bypassed with filter capacitors, and it is recommended that separate PC traces be used to route the 5-V and ground to the reference inputs and to the supply pins.

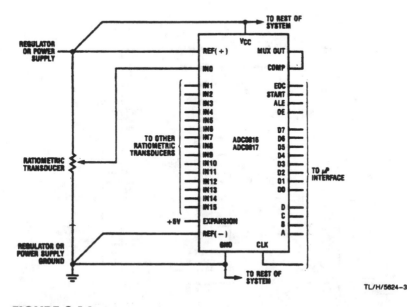

TL/H/5624–3

FIGURE 8-16 Ratiometric conversion with power-supply reference (National Semiconductor, *Linear Applications Handbook,* 1994, p. 592)

8.3.3 Absolute Conversion

Figure 8-17 shows the ADCs connected for simple absolute conversion. *Absolute conversion* refers to the use of transducers with which the output value is not related to another voltage. The "absolute" value of the output voltage of the transducer is very important (in contrast to the output voltage of a ratiometric transducer). This implies that the reference must be accurate to determine the value of the absolute output of the transducer.

In the circuit of Fig. 8-17, a precise, adjustable reference is provided by an LM336-5.0 and the associated parts. Ratiometric transducers also can be used in this circuit and in most of the following applications. However, the key point to remember is that accuracy of absolute conversion depends primarily on the accuracy of the reference voltage. With ratiometric systems, accuracy is determined by the transducer characteristics.

8.3.4 Using the Reference As the Supply

In some small systems (particularly CMOS) in which a reference is required, it is possible to use the reference to provide both the ADC-reference function and regulation for the supply. Figure 8-18 shows such a system in which the LM336-5.0 provides regulated 5-V for the ADC power and reference inputs, as well as for the power inputs of other component in the system. Of course, an unregulated supply greater than 5 V is required for +V.

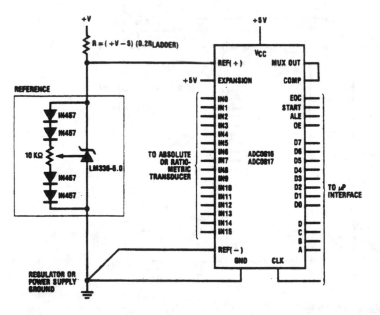

TL/H/5824-4

FIGURE 8-17 Simple absolute conversion (National Semiconductor, *Linear Applications Handbook,* 1994, p. 594)

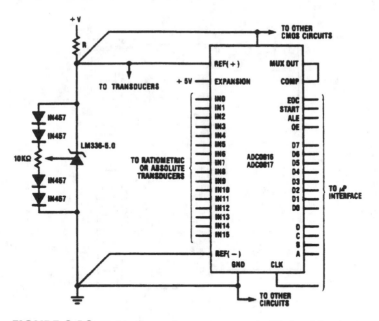

FIGURE 8-18 Reference used as a power supply (National Semiconductor, *Linear Applications Handbook,* 1994, p. 594)

The series resistor R is chosen such that the maximum current needed by the system is supplied, and the LM336-5.0 is kept in regulation. The value of this resistor is found with the following equation:

$$R = (VS - V_{REF})/(ILAD + ITR + IP + IR)$$

where VS = unregulated supply voltage; V_{REF} = reference voltage; ILAD = V_{REF}/lk, resistor ladder current; ITR = transducer currents, IP = system power supply requirements; and IR = minimum reference current.

Figure 8-19 shows a simple method of buffering the references to provide higher current capabilities. This eliminates the IP term in the equation for resistor R. Note that it is advisable to add some supply bypass capacitors (typically 0.1 μF) to reduce noise in the circuits of Figs. 8-18 and 8-19 (as well as Fig. 8-17). The bypass capacitors should be added to the circuits at various supply line points as necessary.

8.3.5 Eliminating Input Gain Adjustments

In some cases, it is possible to eliminate gain adjustments on the analog input signals by varying the ADC REF+ and REF− voltages to get various full-scale ranges. Figures 8-20 and 8-21 show two such circuits. In general, the reference voltage can be varied from 5 V to about 0.5 V to accommodate various input voltages. However, there is one restriction: The center of the reference voltage must be within ±0.1 V of mid-supply.

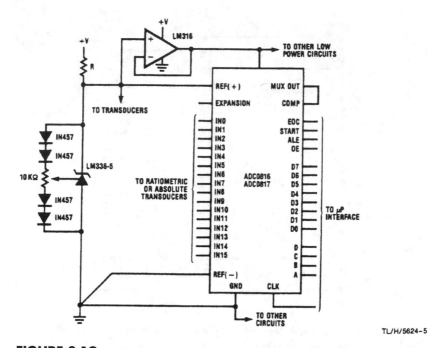

TL/H/5624-5

FIGURE 8-19 Buffered reference used as a power supply (National Semiconductor, *Linear Applications Handbook,* 1994, p. 594)

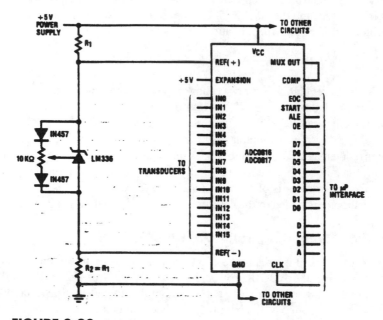

FIGURE 8-20 Supply centered reference with zener (National Semiconductor, *Linear Applications Handbook,* 1994, p. 595)

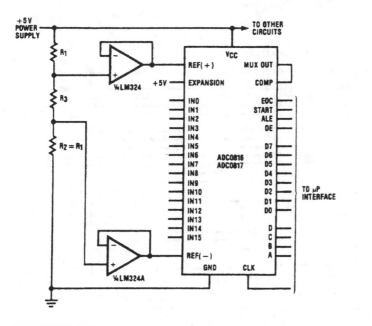

TL/H/5624-6

FIGURE 8-21 Supply centered reference with op amps (National Semiconductor, *Linear Applications Handbook,* 1994, p. 595)

The reason for this restriction is that the reference ladder is tapped by an n-channel or p-channel MOSFET switch tree (Fig. 8-13). Offsetting the voltage at the center of the switch tree from VCC/2 causes the transistors to turn off at the wrong point, resulting in inaccurate and erratic conversions. However, if properly applied, this method can reduce parts counts and eliminate extra power supplies for the input buffers.

In the supply centered reference circuit of Fig. 8-20, R1 and R2 offset REF+ and REF− from VCC and ground. An LM336-2.5 is shown, but any reference between 0.5 V and 5 V can be used.

For odd reference values, use the op-amp circuit of Fig. 8-21. Single-supply op amps such as the LM324 or LM10 can be used. R1, R2, and R3 form a resistor divider in which R1 and R3 center the reference at VCC/2, and R2 can be varied to get the proper reference magnitude.

8.3.6 Expanding Analog Input Channels

Figure 8-22 shows the ADCs connected to provide for 32-channel conversion. Such a configuration is possible because of the EC pin, which is actually a multiplexer enable. When the EC signal is low, all switches are inhibited so that another signal can be applied to the comparator input. Additional channels can be implemented as necessary.

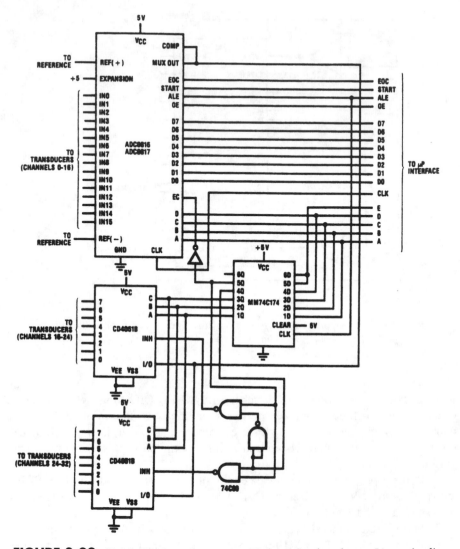

FIGURE 8-22 Simple 32-channel converter (National Semiconductor, *Linear Applications Handbook,* 1994, p. 596)

In the circuit of Fig. 8-22, the number of channels has been expanded from 16 to 32. A total of five address lines are required to address the 32 channels. The lower 4 bits are applied directly to the A, B, C, and D inputs. All 5 bits are also applied to an MM74C174 flip-flop that is used as an address latch for the two CD4051s. The 1Q, 2Q, and 3Q outputs of the flip-flop feed the CD4051 address inputs. The 4Q and 5Q outputs are gated to form enable signals for each CD4051. Output 5Q is also applied at the EC input (after inversion) to enable the ADC multiplexer.

8.3.7 Differential Analog Inputs

Figure 8-23 shows the basic connections for implementing an ADC with differential inputs. With this circuit, the differential inputs are implemented in software. All 16 channels are paired into positive and negative inputs. Then the control logic or microprocessor converts each channel of a differential pair, loads each result, and then subtracts the two results.

This method requires two single-ended conversions to do one differential conversion. As a result, the effective differential-conversion time is twice that of a single channel, or a little more than 200 µs (assuming a clock of 640 kHz). The differential inputs should be stable throughout both conversions to produce accurate results.

Figure 8-24 shows a differential 16-channel converter using one ADC and a few additional parts. This circuit is actually a modified version of the circuit in Fig. 8-22. The CD4051 addressing is changed, and a differential amplifier is added between the multiplexer outputs and the comparator input.

In the circuit of Fig. 8-24, the select logic for the CD4051s is modified to enable the switches so that they can be selected in parallel with the ADC. The outputs of the three multiplexers are connected to a differential amplifier, composed of two inverting amplifiers with gain and offset trimmers. A dual op-amp configuration of inverting amplifiers can be trimmed easily and has less stringent feedback-resistor matching requirements than a single op-amp design.

The transfer equation for the dual op amp shown is:

$$V_{output} = (R2R5)/(R1R3)[V1 - (R5/R4) V2]$$

The propagation delay through the op amps is an important consideration. There must be sufficient time between the analog switch-selection and start-conversion to allow

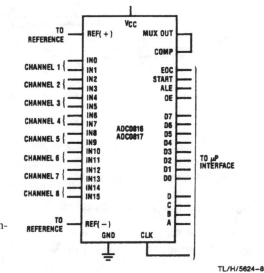

FIGURE 8-23 Basic ADC with differential inputs (National Semiconductor, *Linear Applications Handbook,* 1994, p. 597)

TL/H/5624–8

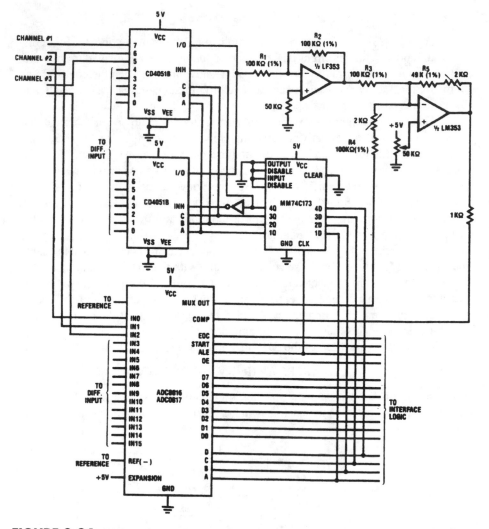

FIGURE 8-24 Differential 16-channel converter (National Semiconductor, *Linear Applications Handbook,* 1994, p. 598)

the analog signal at the comparator input to settle. Using the LF353 op amp shown, the delay is about 5 μs. The op-amp gain and offset controls are adjusted to provide the zero and full-scale digital-output readings for the analog-input range or span.

8.3.8 Buffering Considerations

Figures 8-25 through 8-27 show some typical buffering circuits for the ADCs. Three basic ranges of input signal levels can occur when ADCs are interfaced to the real world. These are as follows: (1) signals that exceed VCC or go below ground, (2) signals with input ranges less than VCC and ground but are different from the

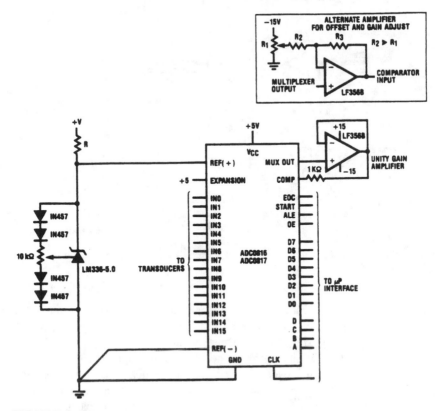

FIGURE 8-25 Single-input buffer (National Semiconductor, *Linear Applications Handbook,* 1994, p. 599)

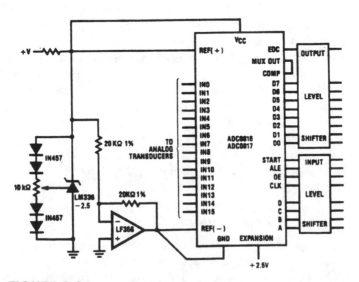

FIGURE 8-26 ±2.5-V input-range data acquisition (National Semiconductor, *Linear Applications Handbook,* 1994, p. 599)

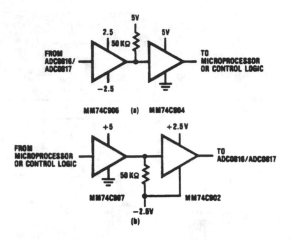

TL/H/5624-11

FIGURE 8-27 Input-output level shifters (National Semiconductor, *Linear Applications Handbook,* 1994, p. 600)

reference range, and (3) signals that have an input range equal to the reference range. Each of these situations requires different buffering.

In the last case (in which the signals are equal to the reference), no buffering is usually required unless the source impedance of the input signal is very high. In this case, a buffer can be added between the multiplexer output and comparator input (see Fig. 8-25). An op amp with high input impedance and low output impedance reduces input leakage (when one views the configuration from the multiplexer).

If the input signal is within the supply range but different from the reference range (or when the reference cannot be manipulated to conform to the full input range), the unity-gain buffer of Fig. 8-25 can be replaced with another op amp as shown in the inset in Fig. 8-25. This type of amplifier provides gain or offset control to produce a full-scale range equal to the reference.

When input range exceeds VCC or goes below ground, the input signals must be level-shifted before the input can go to the multiplexer. There is a limit to such level shifting when the input voltage range is within 5 V but outside the 0.5-V supply range. In this case, the supply for the entire chip can be shifted to the input range, and the digital-output signals can be level-shifted to the system 5-V supply.

A typical example of level-shifting and buffering is the situation in which the bipolar inputs range from −2.5 V to +2.5 V. If the ADCs have the supply and reference provided as shown in Fig. 8-26, then the ±2.5-V logic output can be shifted to 0-V and 5-V logic levels as shown in Fig. 8-27.

8.3.9 Digital Data Acquisition

Although the ADCs are designed for analog data acquisition, they also can be used for digital data acquisition in some cases. For example, if a system has unused channels, digital inputs can be connected to these channels instead of being separately buffered into the system. In the case of a microprocessor system, this could eliminate an I/O port and associated logic. The speed at which the inputs are accessed is one conversion cycle (which is fast enough for many applications).

The ADC inputs also can be used as input switches, power-supply indicator devices, or other system-status flags. In any configuration, the microprocessor converts the digital input channel and reads it. Software then decides whether the input is high enough (or low enough) to cause a particular action.

8.3.10 Input Considerations

Figure 8-28 shows a simplified model of the multiplexer-comparator input. As shown, the input is a sampling-type comparator in which the input current is a series of spikes, not a small DC current as might be expected. This can present a problem in some applications. The following are some points to consider.

When determining a single-bit value during a conversion, S1 closes, causing CC and CP to charge to the input voltage. Then S1 is opened and S2 is closed to sample the ladder. The resulting input current is an RC-transient charging current in which magnitude and duration depend on the values of CC, CP, RS, RM, and RL.

The duration of the transient current must be shorter than the input-sample period. In turn, the sample period depends on the converter clock frequency. As a result, the maximum source impedance (RS) depends on the clock period. As a guideline, at a clock frequency of 1 MHz the source impedance is equal to or less than 1 k. When the clock is reduced to 640 kHz, the source impedance increases to 2 k (or less). The source impedance of potentiometric transducers varies as a function of wiper position. Thus transducers with a value of 10 k (or less) are suitable for clock frequencies of 640 kHz. When the clock is increased to 1 MHz, the transducer should have a source impedance of 5 k (or less).

When a sample-and-hold or other active device (such as an op amp) is inserted between the multiplexer and comparator pins (as in Fig. 8-25), the output impedances of the transducers are no longer critical. Instead, the impedances of the sample-and-

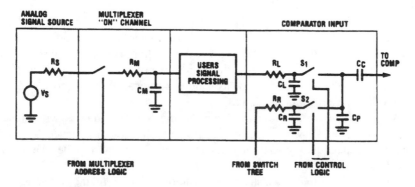

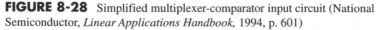

FIGURE 8-28 Simplified multiplexer-comparator input circuit (National Semiconductor, *Linear Applications Handbook,* 1994, p. 601)

hold or op amp become the determining factor. At a clock frequency of 1 MHz, the buffer impedance should be 3 k (or less). With the clock at 640 kHz, the impedance can be 5 k (or less).

If it is not possible to avoid higher source impedances than these recommended guidelines, RC charging errors can be reduced to an average current error by placing a 1-μF capacitor from the multiplexer input to ground. Adding the capacitor averages the transient-current spikes and causes a small DC error.

For a potentiometric transducer, the DC error is:

$$V_{ERR} = (RP/8) (IN) (CK/640 \text{ kHz}) \text{ volts}$$

where RP = total potentiometric resistance; IN = 2 μA (maximum input current at 640 kHz); and CK = clock frequency.

For a standard buffer source impedance, the DC error is:

$$V_{ERR} = (IN)(RS) (CK/640 \text{ kHz}) \text{ volts}$$

where IN = 2 μA (maximum input current at 640 kHz); RS = buffer source impedance; and CK = clock frequency.

In addition to source-impedance considerations, several precautions should be observed whenever analog signals are present in a digital system. To reduce noise on the analog inputs, keep the reference input and the analog input signals physically isolated from any digital signals. As always, use a single-point ground.

8.3.11 Analog Input Overvoltage Protection

Figure 8-29 shows some protection circuits for the analog inputs. For proper operation, it is important to keep the analog input voltages to the multiplexer or comparator between VCC and ground. Of course, there might be instances in which overvoltage or undervoltage cannot be avoided. Protecting analog inputs, because of their unique nature, is usually more difficult than protecting digital inputs (which are typically on or off).

As shown in Fig. 8-29, the most effective analog-input overvoltage production circuits use some combination of Schottky diodes. Because the Schottky knee voltage is 0.4 V, the 1N5168 diodes of Fig. 8-29a safely shunt currents up to several milliamperes. For larger currents, series resistances can be added to limit current as shown in Fig. 8-29b. Do not use resistor values greater than those shown or higher than the values described in Section 8.3.10.

A less expensive solution is to replace the Schottky diodes with standard switching diodes as shown in Fig. 8-29c. However, a series resistance is required because standard diodes only partially shunt the input current (from the clamp diodes within the IC).

If the external diode must shunt a large amount of current, the two series resistors of Fig. 8-29d should be used. If the design is such that the input can exceed only one supply, the diode going to the other supply can be omitted.

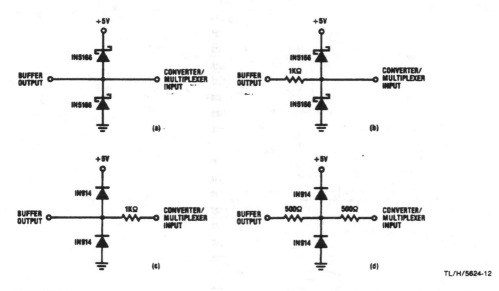

FIGURE 8-29 Protection circuits for analog inputs (National Semiconductor, *Linear Applications Handbook,* 1994, p. 601)

8.3.12 Signal Conditioning

There are many applications in which it is desirable to add signal-processing circuits to improve converter performance. Typical additions are filter circuits, sample-and-holds, and gain-controlled amplifiers. Here again, the external accessibility of the multiplexer-output and comparator-input pins can greatly reduce the need for external circuits. This is because only one circuit is required by all 16 outputs, instead of one for each input. The following paragraphs describe some typical signal-conditioning applications.

8.3.13 Microprocessor-Controlled Gain

Figure 8-30 shows the ADCs connected for an external gain-control under supervision of a microprocessor. The CD4051 analog multiplexer is placed in the feedback loop of a simple non-inverting op amp. One controls the gain of this op amp by selecting one of the CD4051 analog switches. This switches a resistor in and out of the feedback loop. If the resistors R2N are of different value, different gains are realized. The gains are given by: Gain (AV) = 1 + (R2N/R1). A microprocessor (or some control logic) selects a gain by latching the channel address into an MM74C173.

It is important that the LF356B output not exceed the power supply, so, before a new channel is selected, the op-amp gain must be reduced to a new level. The 1-k resistor at the LF356B output helps protect the comparator inputs from accidental overvoltage (or undervoltage). The two back-biased diodes at the input to VCC and ground (1N914 or Schottky) offer further protection.

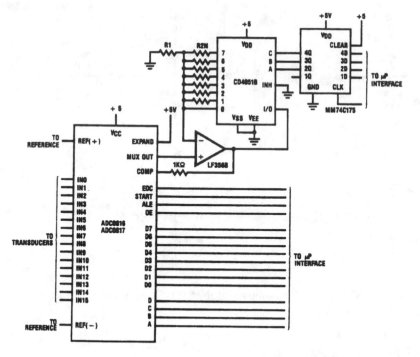

TL/H/5624-13

FIGURE 8-30 ADC with microprocessor-controlled gain (National Semiconductor, *Linear Applications Handbook*, 1994, p. 602)

8.3.14 Sample-and-Hold

Figures 8-31 and 8-32 show the ADCs connected for sample-and-hold (S/H) operation. The S/H function is the only major data-acquisition element not included in these ADCs. If the input signals are fast moving, then an S/H should be used to quickly acquire the signal and then hold the signal while the ADCs convert it to a digital readout. This can be easily implemented by means of insertion of the S/H between the multiplexer output and the comparator input as shown.

In the simplest form, the multiplexer output is connected to the comparator input, with a capacitor connected to ground (similar to that shown in Fig. 8-31 but without the op amp). The expansion-control pin is used as a sample-control input. When EXPAND is high, one switch is on and the capacitor voltage follows the input. When EXPAND is low, all switches are turned off and the capacitor holds the last value.

Unfortunately, this simple solution is not practical. The input bias to the comparator is about 2 μA (worst case, with a clock of 640 kHz). The droop or discharge rate for a 1000-pF capacitor is about 2000 V/s or about 0.2-V per conversion. This is not practical. If a 0.01-μF capacitor is used instead, the rate is about 20 μV, which might work. However, the acquisition time would be about 100 μs, or about the length of a conversion.

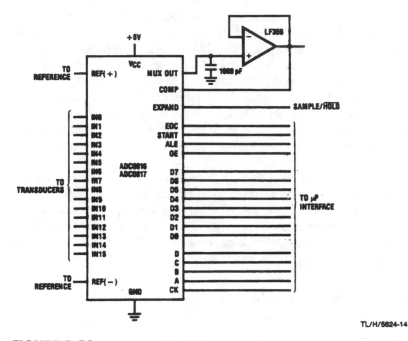

TL/H/5624-14

FIGURE 8-31 Sample-and-hold with op amp (National Semiconductor, *Linear Applications Handbook,* 1994, p. 603)

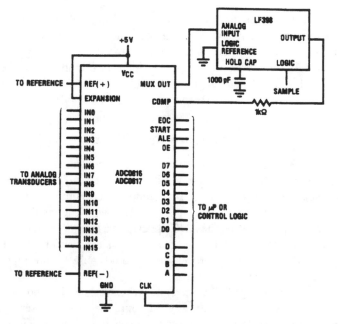

TL/H/5624-15

FIGURE 8-32 Sample-and-hold with IC S/H (National Semiconductor, *Linear Applications Handbook,* 1994, p. 604)

The circuit of Fig. 8-31 eliminates the problem produced by the high comparator-input leakage. With the LF356 buffer connected between the multiplexer-output and comparator-input pins, the leakage is reduced from 2 μA to about 100 nA. The droop-per-conversion is typically less than 1.0 μV per conversion (with a 1000 pF capacitor) and the acquisition time is about 20 μs (instead of the 100 μs).

The circuit of Fig. 8-32 isolates the capacitor from both the multiplexer and comparator pins using an LF398 IC S/H. Acquisition time for the LF398 is a typical 4 μs (to 0.1%), and the droop rate is about 20 μV/conversion. Because the LF398 has its own S/H input, the expansion control of the ADC is free to be used in the normal manner.

No matter what configuration is used, the choice of the hold capacitor is critical to the performance of the S/H circuit. Some capacitors are composed of dielectrics that have an initial droop after the hold function is strobed. This is because of dielectric absorption. Polypropylene and polystyrene dielectrics have very little dielectric absorption and thus make good S/H capacitors. Materials such as Mylar polyethylene have higher absorption properties and should not be used for S/H circuits.

8.3.15 Filtering Analog Inputs

Filtering might be required for some analog-input signals, especially signals from a noisy environment. High-frequency noise is a particular problem. The ADCs can accommodate the addition of most standard low-pass filters. Another useful filter is a 50-Hz or 60-Hz notch filter to eliminate the noise contributed to the circuit by AC power lines.

A single passive filter can be placed between the multiplexer-output and comparator-input pins. However, such passive filters must be carefully designed to reduce input loading. The filter capacitor tends to average the comparator sampling current. To eliminate this effect, use an op amp to buffer the filter, or use an active filter. If you are not familiar with filter-circuit design, read *McGraw-Hill Circuit Encyclopedia & Troubleshooting Guide* volumes 1, 2, and 3 (McGraw-Hill, 1996).

8.3.16 Microprocessor Interface Considerations

Figures 8-33 and 8-34 are flow charts for interrupt-control and polled-I/O modes of microprocessor interface, respectively. Either interface can be used, but the polled-I/O method usually requires fewer external components. With polled-I/O, the microprocessor or CPU periodically interrogates the ADC, which looks like an I/O port to the CPU. With interrupt-control, the ADC appears as a memory and interrupts the CPU. From a simplified-design standpoint, the major concern is whether the EOC (Fig. 8-15) should be polled by the microprocessor or should cause an interrupt. The remainder of this section describes both approaches.

Even though the actual timing of CPU read and write cycles varies, most microprocessors output the address and data (during write) onto the system buses. A certain time later, the read or write strobes go active for a specified time. The interface logic must detect the state of the address and data buses and initiate the action. For

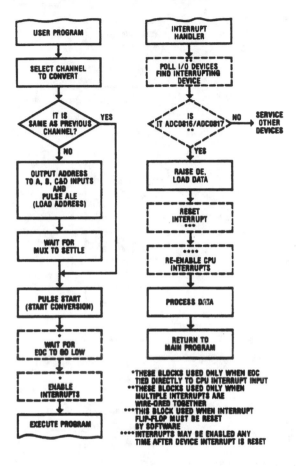

FIGURE 8-33 Flow chart for interrupt control (National Semiconductor, *Linear Applications Handbook,* 1994, p. 605)

the ADCs in this chapter, these actions are: (1) load channel address, (2) start conversion, (3) detect EOC, and (4) read the resultant data. One performs these actions by decoding the read-write strobes, address, and data to form ALE and START pulses, then to detect EOC, and finally to read the data.

Typical decoding and strobe generation are straightforward. The START, ALE, and OUTPUT ENABLE strobes generally are of the same duration as the CPU read-write strobes and are positive-going (although ALE can be negative-going). One possible concern is where to take the A, B, C, and D channel-select address. These lines can be connected to either the address bus or the data bus. The advantage of using the data bus is that in minimum systems, more I/O address lines are available for simple decoding. When the A, B, C, and D inputs are connected to the address bus, each analog channel becomes a separate I/O port.

It is possible to connected the START and ALE pins together so that one pulse can both write the channel address and start the conversion. However, it is essential that the comparator-input signal be stable before conversion starts. If not, the first (and

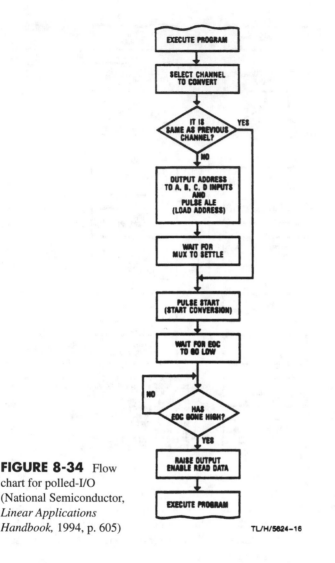

FIGURE 8-34 Flow chart for polled-I/O (National Semiconductor, *Linear Applications Handbook,* 1994, p. 605)

TL/H/5624–16

most important) successive approximation could be in error. Typical START and ALE pulses are the same length as the CPU read and write strobes (normally 0.2- to 1-μs long). As a result, conversion can start within 1 μs of the address-select latching. (As shown in Fig. 8-15, the channel is selected on the rising edge of ALE, and conversion begins within eight clock periods of the falling edge of START.)

When clock speed is greater than 500 kHz, 1 μs might not be enough time to allow the analog input signal to pass through the multiplexer and any additional signal-conditioning circuits such as buffers and S/H. However, there are two relatively simple ways to overcome this problem. First, the START/ALE pulse can be stretched to the desired length with a one-shot (such as an MM74C221). This produces a pulse as long as the total delay from multiplexer input to comparator input.

As an alternative, the converter can be "double pulsed" in software by writing to the START/ALE address twice. The first pulse latches the desired channel address and starts the conversion. The second pulse must again load the same channel address, which does not change the multiplexer state, and then restarts the conversion. The second pulse must occur after the comparator input has settled.

8.3.17 Interfacing 8080-Type Microprocessors

Figures 8-35 and 8-36 show connections for interface to 8080-type micro-processors (INS8080/8224/8228). This interfacing is quite simple because the INS8080 CPU has separate I/O read (I/OR) and I/O write (I/OW) strobes (or separate I/O addressing). As a result, in these simple interface systems, little or no address decoding is required.

As shown in Fig. 8-35, two NOR gates are used to gate the I/O strobes with the most-significant address bit A7. (The INS8080 has 8 bits of port address, yielding a maximum of four I/O ports if inputs A, B, C, and D are connected to the address bus.) An MM74C74 flip-flop is used as a divide-by-2 to generate a converter clock of 1 MHz. If the system clock is equal to or less than 1 MHz, the flip-flop can be omitted.

Typical software for the Fig. 8-35 circuit first writes the channel address to the converter as a start signal. Two start pulses are sent to the ADCs to allow the

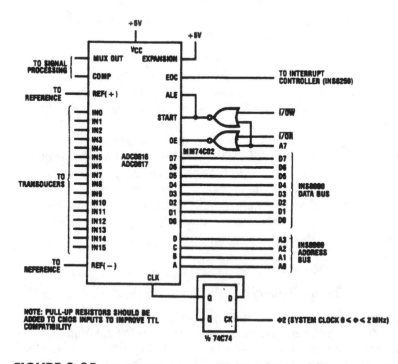

FIGURE 8-35 Simple 8080 interface (National Semiconductor, *Linear Applications Handbook,* 1994, p. 606)

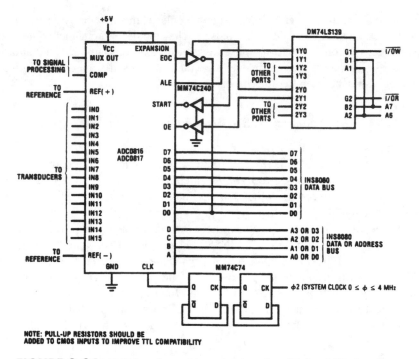

FIGURE 8-36 8080 interface with partial decoding (National Semiconductor, *Linear Applications Handbook,* 1994, p. 607)

comparator input to settle. After the second start pulse, the CPU can execute other program segments until the CPU is interrupted by EOC going high. Depending on interrupt structure, program control then is given to the interrupt handler, which reads the converter data.

The circuit of Fig. 8-36 uses a DM74LS139 dual 2–4 decoder in which one-half of the chip is used to create read pulses, and the other half to create write pulses. The START and OE inputs are inverted to provide the correct pulse polarity. This interface partially decodes A6 and A7 to provide more I/O capabilities than the Fig. 8-35 circuit. The Fig. 8-36 circuit also implements a simple polled-I/O structure. The EOC output is placed on the data bus by a tristate inverter when the inverter is enabled by a read pulse from the INS8080.

8.3.18 Interfacing Z80-type Microprocessors

Figures 8-37 and 8-38 show connections for interface to Z80-type microprocessors. The Z80, even though architecturally similar to the INS8080, uses slightly different control lines to perform I/O reads and writes. In Fig. 8-37, NOR gates are used to strobe the I/O functions. However, the Z80 has RD (read) and WR (write) strobes, which are gated with IOREQ (I/O request).

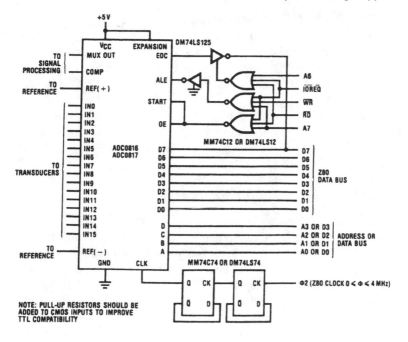

TL/H/5624-18

FIGURE 8-37 Simple Z80 interface (National Semiconductor, *Linear Applications Handbook*, 1994, p. 607)

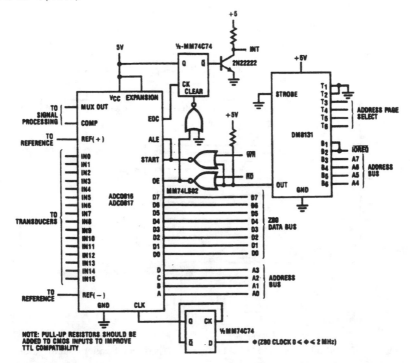
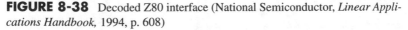

FIGURE 8-38 Decoded Z80 interface (National Semiconductor, *Linear Applications Handbook*, 1994, p. 608)

In the circuit of Fig. 8-37, START is connected to OE. This causes a new conversion to be started whenever the data bits are read. This might seem unusual, but it can be useful if the converter is to be continually restarted on completion of the previous conversion. Address bit A6 is used to drive a strobe that places EOC on the data bus to be read by the CPU.

In the circuit of Fig. 8-38, a 6-bit comparator is used to decode A4-A7 and IOREQ. Two NOR gates are used to gate the ALE/START and OE pulses. This design functions the same as that of Fig. 8-37, except that the DM8131 provides much more decoding.

8.3.19 Interfacing NSC800 Microprocessors

Figures 8-39 and 8-40 show connections for interface to NSC800 microprocessors. This interface is quite similar to that for the 8080, even though the timing is very different. The NSC800 multiplexes the lower 8 address bits on the data bus at the beginning of each cycle. When accessing memory, A0-A7 must be latched out at the beginning of a read or write cycle.

For I/O accessing, the NSC800 duplicates the 8-bit I/O addresses on A8-A15 address lines. Latches are not necessary because these lines are not multiplexed. The I/O read and write strobes are taken from RD (read) and WR (write) lines and the IO/M signal.

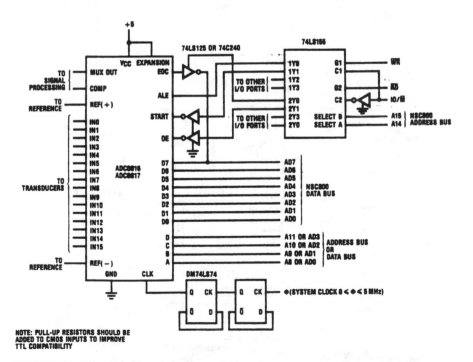

FIGURE 8-39 Partially decoded NSC800 interface (National Semiconductor, *Linear Applications Handbook,* 1994, p. 608)

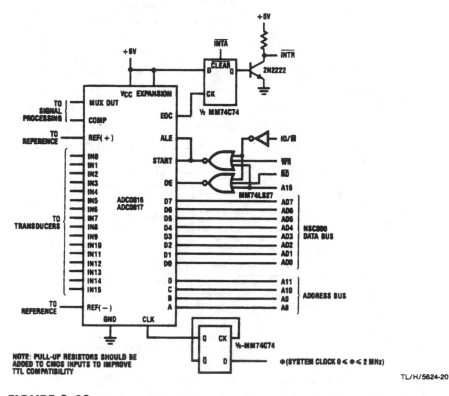

FIGURE 8-40 Minimum NAC800 interface (National Semiconductor, *Linear Applications Handbook,* 1994, p. 609)

The circuit of Fig. 8-39 uses a dual 2–4 line decoder that decodes A15. A14 is enabled by the read-write strobes. Tristate inverters are used to implement a decoding similar to that of Fig. 8-36. Double pulsing is not required because START and ALE are accessed separately.

The circuit of Fig. 8-40 uses NOR gates (similar to that of Fig. 8-35) but with different control signals. When EOC goes high, the flip-flop is set, and INTR goes low. When the NSC800 acknowledges the interrupt by lowering INTA, the flip-flop resets. If more than one interrupt can occur simultaneously, either INTA should be gated with EOC, or a signal other than INTA must be used. This is required because the NSC800 can detect another interrupt and clear the ADC interrupt before the ADC signal is detected.

8.3.20 Interfacing 6800-type Microprocessors

Figures 8-41 and 8-42 show connections for interface to classic 6800 microprocessors. Because the 6800 has no separate I/O addressing capabilities, the system I/O must be addressed as though it is memory. As discussed, memory mapping can

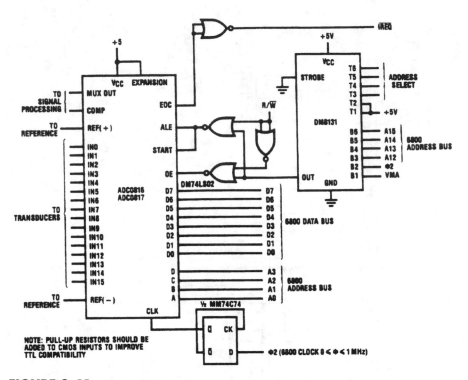

FIGURE 8-41 Simple 6800 interface (National Semiconductor, *Linear Applications Handbook,* 1994, p. 610)

require more address decoding to separate memory for I/O. In small systems, however, the parts count can still be kept to a minimum.

Figure 8-41 shows an interface in which a DM8131 comparator is used to partially decode the A12, A13, A14, and A15 address lines with the 2 clock and VMA (valid memory address). This provides an address decode pulse for the two NOR gates, which in turn generate the START/ALE pulse and the output-enable OE signal. The design locates the ADC in one 4-k byte or block.

The circuit of Fig. 8-41 ties EOC to IREQ interrupt through an inverter, and is usable only in single-interrupt systems because the 6800 has no way of resetting the interrupt (except by starting a new conversion). Because EOC is directly tied to the interrupt input, the controlling software must not re-enable interrupts until eight converter clock periods after the START pulse, when EOC is low.

Figure 8-42 shows an interface with more I/O port strobes. A NAND gate and inverter are used to decode the addresses, VMA, and 2 clock. The I/O addresses are located at 11110XXXXXAABBBB (binary); where X = don't care; A = 00 (binary) for ALE write or IREQ reset/EOC read and A = 01 for START write or data read; and B = channel-select address, if A, B, C, and D are connected to the address bus and ALE is accessed. A dual 2–4 line decoder is used to generate these strobes. Inverters are used to create the correct logic levels.

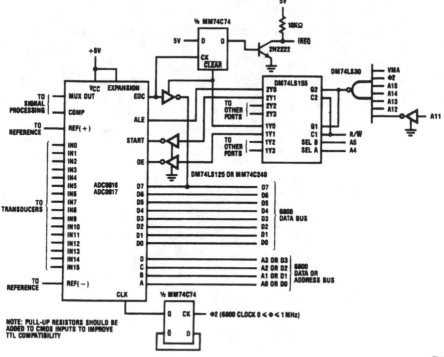

TL/H/5624-21

FIGURE 8-42 Partially decoded 6800 interface (National Semiconductor, *Linear Applications Handbook,* 1994, p. 610)

The 6800 supports only a wired-OR interrupt structure. In a multi-interrupt environment, only one interrupt is received and the interrupt-handler routine must determine which device has caused the interrupt and must service that device. To do this, the EOC is brought out to the data bus so that EOC can be checked by the CPU.

8.3.21 Parallel Interface

In some cases, microprocessor-support ICs can be used to interface the ADCs to microprocessors. Most parallel I/O chips can be used, and they provide enough flexibility for all functions to be under software control.

The manufacturer recommends INS8255, 6820, and Z80s-PI0 (or some similar device) for parallel I/O. In most cases, these support ICs can be connected directly to the data and control pins. Software is then used to manipulate the START and ALE pins via the interface chip. In some cases, the chips provide handshaking or interrupt capabilities and possibly a clock. If not, such functions must be provided by other circuits. Keep in mind that parallel I/O circuits are usually more expensive than the circuits they replace, although such support ICs can simplify design and increase versatility.

8.4 Circuit Applications with Multiplying DACs

Figure 8-43 shows the internal functions of a multiplying DAC. Because such four-quadrant DACs allow a digital word to operate on an analog input, or vice versa, the output can represent a sophisticated function. CMOS multiplying DACs allow true bipolar analog signals to be applied to the reference input. This feature makes such DACs useful in many applications not generally considered data converters. The following paragraphs describe some typical applications for multiplying DACs.

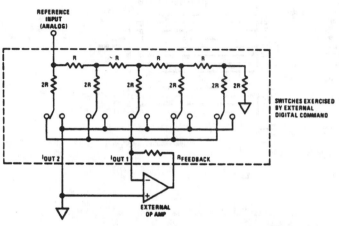

Details (Simplified) of CMOS DAC1020—Last 5 Bits Shown

Other CMOS DACs are similar in the nature of operation but also include internal logic for ease of interface to microprocessor based systems. Typical is the DAC1000 shown below.

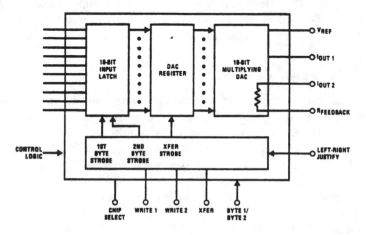

TL/H/5628-1

FIGURE 8-43 Internal functions of a multiplying DAC (National Semiconductor, *Linear Applications Handbook,* 1994, p. 659)

8.4.1 Sine-Wave Generator with Digital Control

Figure 8-44 shows a variable-frequency sine-wave generator capable of producing signals at frequencies up to 30 kHz under digital control. This function is valuable in automatic test equipment and instrumentation applications and is not easily achieved in normal sine-wave generation circuits. The linearity of output frequency to digital-code input is within 0.1% for each of the 1,024 discrete output frequencies.

To understand circuit operation, assume that the A2 output is negative. This means that the zener-determined output of A2 applies -7 V to the DAC reference input. Under these conditions, the DAC pulls a current from the A1 summing junction. The current is directly proportional to the digital code applied to the DAC. Integrator A1 ramps far enough so that the potential at the A2 +input just goes positive. The A2 output changes state, and the potential at the DAC reference input changes to +7 V. The DAC output current reverses and the A1 integrator is forced to move in the negative direction. When the negative-going output of A1 becomes large enough to pull the A2 +input slightly negative, the A2 output changes state and the process repeats.

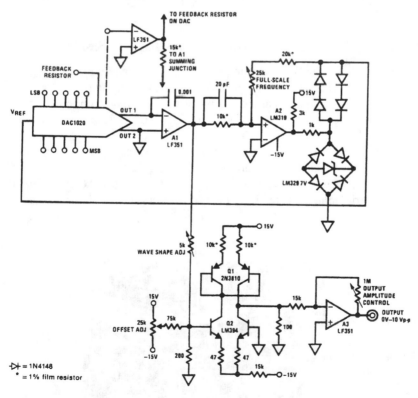

TL/H/5628–2

FIGURE 8-44 Sine-wave generator with digital control (National Semiconductor, *Linear Applications Handbook,* 1994, p. 660)

The amplitude-stabilized triangle wave at the A1 output has a frequency that depends on the digital word at the DAC. The 20-pF capacitor provides a slight leading response at high frequencies to offset the 80-ns response time of A2. This aids overall circuit linearity. The triangle wave is applied to the Q1/Q2 shaper network, which provides a sine-wave output.

To adjust the circuit, set all DAC digital inputs high and trim the 25-k pot for 30-kHz output (using a frequency counter). Then connect a distortion analyzer to the circuit output and adjust the 5-k and 75-k pots for minimum distortion. Finally, set the 1-M output control for the desired output.

This circuit provides rapid switching of output frequency, as shown in Fig. 8-45. Note that the output frequency shifts immediately (actually with no undesired delay) by more than an order of magnitude in response to digital commands (top line of Fig. 8-45).

If operation over temperature is required, the absolute change in resistance in the DAC internal ladder might cause unacceptable errors. This can be corrected by reversing the A2 inputs and inserting an amplifier (dashed lines in Fig. 8-44) between the DAC and A1. Because this amplifier uses the DAC internal feedback resistor (Fig. 8-43), the temperature error in the ladder is cancelled. This results in more stable operation.

8.4.2 Digitally Programmable Pulse-Width Modulation

Figure 8-46 shows the DAC used to program the pulse width of clock signals under digital control. This function is valuable in automatic testing of secondary-breakdown limits for switching transistors. However, the high resolution of control that the circuit has over the pulse width is useful anywhere wide range, precision, pulse-width modulation (PWM) is required.

In this circuit, the length of time required by the A1 integrator to charge to a reference level is determined by the current from the DAC. In turn, the current is directly proportional to the digital code at the DAC input. Both the DAC analog input and the reference trip point are taken from the LM329 voltage reference.

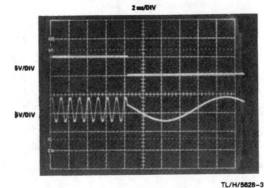

FIGURE 8-45 Sine-wave generator oscilloscope display (National Semiconductor, *Linear Applications Handbook,* 1994, p. 660)

TL/H/5628–3

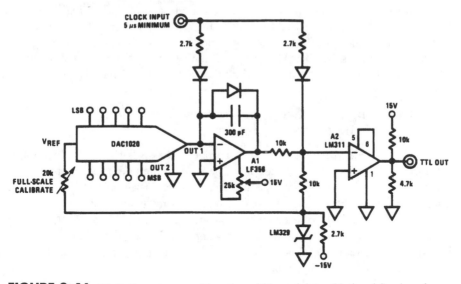

FIGURE 8-46 Digitally programmable pulse-width modulator (National Semiconductor, *Linear Applications Handbook,* 1994, p. 661)

While the integrator output (trace A of Fig. 8-47) is below the trip point, A2 comparator output remains high (trace B). When the trip point is exceeded, A2 output goes low. The DAC input code can vary the output pulse width over a range determined by the DAC resolution.

Traces C, D, and E of Fig. 8-47 show the fine detail of the resetting sequence. (The horizontal scale is expanded for these traces.) Trace C is the 5-μs clock pulse. When this pulse rises, the A1 integrator output (trace D) is forced negative until the output reaches the limit of the diode in the feedback loop. While the clock pulse is high, current through the 2.7-k diode path forces the A2 output low. When the clock pulse goes low, the A2 output goes high and remains high until the A1 integrator output-amplitude exceeds the trip point.

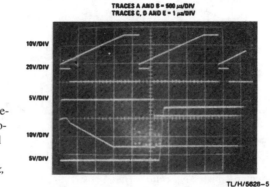

FIGURE 8-47 Pulse-width modulator oscilloscope display (National Semiconductor, *Linear Applications Handbook,* 1994, p. 661)

TL/H/5628–5

To calibrate this circuit, set all DAC bits high, and adjust the FULL-SCALE CALIBRATE pot for the desired full-scale pulse width. Then set only the DAC LSB high, and adjust the A1 offset pot for the appropriate length. (This is 1/1024 of the full-scale value for a 10-bit DAC.) If the 2.2-mV/°C drift of the clamp diode in the A1 feedback loop is objectionable, replace the diode with an FET switch.

8.4.3 Log Amplifier with Digitally Controlled Scale Factor

Figure 8-48 shows the DAC used to program the scale factor of a logarithmic amplifier under digital control. Log amplifiers are commonly used in applications that require a wide dynamic-measurement range (such as in photometry), and it is often necessary to set the amplifier scale factor for some given value.

In this circuit, Q1 is the actual log-converter transistor. Q2 and the 1-k resistor provide temperature compensation. The log-amplifier output is taken at A3. The digital code applied to the DAC determines the overall scale factor of the input-voltage (or current) to output-voltage ratio.

8.4.4 Amplifier with Digitally Controlled Gain

Figure 8-49 shows the DAC used to program the gain of an amplifier under digital control. The amplifier handles bipolar input signals. In this circuit, the input is applied to the amplifier through the DAC feedback resistor. The digital code selected at the DAC determines the ratio between the DAC feedback resistor and the impedance that the DAC ladder presents to the op-amp feedback path.

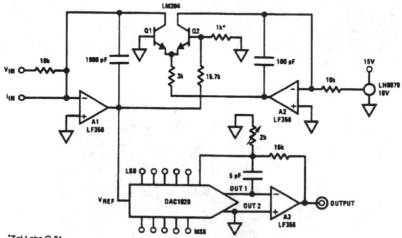

FIGURE 8-48 Log amplifier with digitally controlled scale factor (National Semiconductor, *Linear Applications Handbook,* 1994, p. 662)

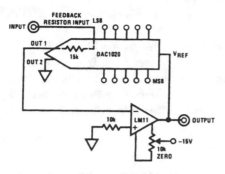

TL/H/5628–7

FIGURE 8-49 Amplifier with digitally controlled gain (National Semiconductor, *Linear Applications Handbook,* 1994, p. 662)

If no digital code (all 0s) is applied to the DAC, there will be no feedback and the amplifier will saturate. If this condition is objectionable, a large-value (22-M) resistor can be shunted across the DAC feedback path. This has little effect at lower frequencies. The gain accuracy of this circuit depends directly on the open-loop gain of the amplifier, not on the DAC.

8.4.5 Digitally Controlled Filter

Figure 8-50 shows the DAC used to program the cutoff frequency of a filter. (The equation given in the figure governs the cutoff frequency.) In this circuit, the DAC provides high-resolution digital control of frequency response by effectively

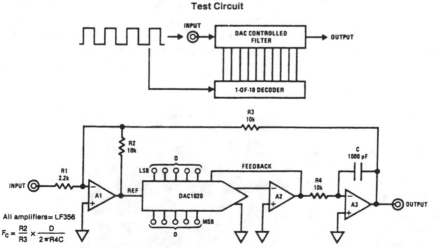

TL/H/5628–8

FIGURE 8-50 Digitally controlled filter (National Semiconductor, *Linear Applications Handbook,* 1994, p. 663)

varying the time constant of the A3 integrator. This is shown in Fig. 8-51 (which is a scope presentation of the test circuit in Fig. 8-50).

When each input square wave is presented to the filter, the 1-of-10 decoder shifts a 1 to the next DAC digital-input line, in sequence. Trace A is the input waveform, and trace B is the waveform at the A1 output (which is the reference input of the DAC). The circuit output at A3 appears at trace C. When the circuit shifts the 1 toward the lower-order DAC inputs, the cutoff frequency decays rapidly (as shown in trace C).

8.5 Some Classic CMOS DAC Applications

This section is devoted to some classic applications for CMOS DACs. Although the circuits show the National Semiconductor MICRO-DAC™ devices, the circuits can be adapted to other DACs, provided that the DACs have similar characteristics and features (such as an internal feedback resistor), and are CMOS. MICRO-DAC™ is a registered trademark of National Semiconductor Corporation.

8.5.1 Digital Potentiometer

Figure 8-52 shows the basic connections for a DAC operated as a digital potentiometer. In this circuit, the applied digital-input word multiplies the applied reference voltage. The resultant output voltage is the product of this multiplication, normalized to the resolution of the DAC. The op amp converts the DAC output current to a voltage through the 15-k feedback resistor within the DAC.

In this particular DAC, the output current ranges from a near-zero output leakage (about 15 nA) for an applied code of all 0s ($D = 0$), to a full-scale value ($D = 2^{n-1}$); where D = decimal equivalent of the binary input, and n = the bits of resolution of the DAC, of V_{REF} divided by the R value of the internal R-2R ladder

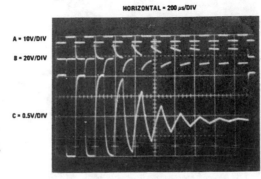

HORIZONTAL = 200 μs/DIV

A = 10V/DIV

B = 20V/DIV

C = 0.5V/DIV

FIGURE 8-51 Digitally controlled filter oscilloscope displays (National Semiconductor, *Linear Applications Handbook*, 1994, p. 663)

TL/H/5628–9

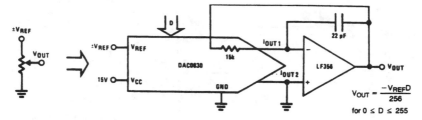

FIGURE 8-52 Digital potentiometer (National Semiconductor, *Linear Applications Handbook,* 1994, p. 665)

network (nominally 15 k). The current at IOUT 2 is equal to that caused by the one's-complement of the applied digital input (with IOUT 1 at full-scale, IOUT 2 is zero).

The output voltage is the opposite polarity of the applied reference voltage. However, because the DAC is CMOS, bipolar reference voltages can be used. For example, if a positive output is needed, a negative reference can be applied.

To preserve output linearity, the two current-output pins must be as close to 0 V as possible. This requires that the input-offset voltage of the op amp be nulled. The amount of linearity-error degradation is about $V_{OS} + V_{REF}$. When the digital pot is used to attenuate AC signals (in audio applications for example), the DAC linearity over the full range of the applied reference voltage (even if it passes through zero) is good enough to distort a 10-V sine wave by only 0.004%.

The feedback capacitor shown in Fig. 8-52 is added to improve the settling time of the output when the input code is changed. Without this compensation, there can be ringing and overshoot in the output (because of the usual feedback-pole caused by the feedback resistor and the DAC output capacitance).

As usual, the op amp should have good DC characteristics (low offset voltage (VOS) and low VOS drift) as well as fast AC characteristics (high slew rate, short settling time, and wide bandwidth). If any of these terms is unfamiliar, read my *Simplified Design of IC Amplifiers* (Newnes, 1996).

If it is not practical to find an op amp that meets all these requirements, it is possible to use a combination of op amps such as shown in Fig. 8-53. This circuit combines the excellent DC input characteristics of the classic LM11 with the fast response of an LF351 (a combination bipolar device).

8.5.2 Level-Shifting the Output Range

Figure 8-54 shows the basic connections for a DAC operated to shift the output level. The shift is made by summing a fixed current to the DAC current-output terminal, offsetting the output voltage to the op amp. The applied referenced voltage then serves as the output-span controller and is added (in fractions) to the output as a function of the applied digital code.

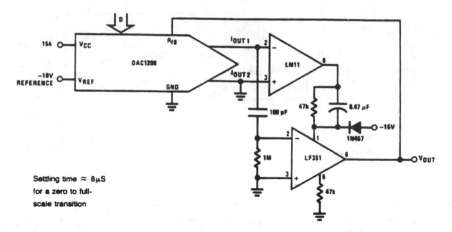

FIGURE 8-53 Digital potentiometer with composite amplifier (National Semiconductor, *Linear Applications Handbook,* 1994, p. 665)

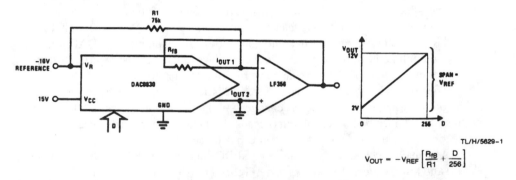

FIGURE 8-54 DAC with level-shifted output (National Semiconductor, *Linear Applications Handbook,* 1994, p. 665)

8.5.3 Single-Supply Operation

Figure 8-55 shows the DAC connected for single-supply operation. The R-2R ladder can be operated as a voltage-switching network to circumvent the output-voltage inversion inherent in the current-switching mode. In this circuit, the reference voltage is applied to the IOUT 1 terminal and is attenuated by the R-2R ladder in proportion to the applied code. The voltage is then output to the V_{REF} terminal with no phase inversion.

To ensure linear operation in this mode, the applied reference voltage must be kept less than 3 V for 10-bit DACs, or less than 5 V for 8-bit DACs. The supply voltage to the DAC must be at least 10 V more positive than the reference voltage to ensure that the CMOS ladder switches have enough voltage overdrive to fully turn

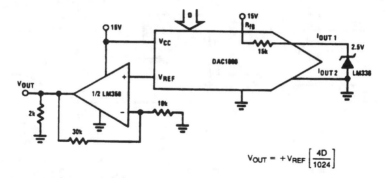

$$V_{OUT} = + V_{REF} \left[\frac{4D}{1024} \right]$$

FIGURE 8-55 DAC connected for single-supply operation (National Semiconductor, *Linear Applications Handbook,* 1994, p. 666)

on. An external op amp can be added to provide gain to the DAC output voltage for a wide overall output span.

The zero-code output voltage is limited by the low-level output-saturation voltage of the op amp. The 2-k load resistor helps to minimize the voltage. This circuit provides generally good linearity for 8-bit and 10-bit DACs but can have a linearity problem with 12-bit DACs (because of the very low reference required). DACs designed specifically for single-supply operation (including 12-bit DACs) are described in this chapter. Such DACs should be used if single-supply operation is essential.

8.5.4 Bipolar Output from a Fixed Reference Voltage

Figure 8-56 shows the DAC connected to provide a bipolar output from a fixed reference voltage. This connection is made with a second op amp in the analog-output circuit. In effect, the circuit gives sign significance to the MSB of the digital-input word, allowing four-quadrant multiplication of the reference voltage. The polarity of the reference can still be reversed (or can be an AC signal) to realize full four-quadrant multiplication.

8.5.5 DAC-Controlled Amplifier

Figure 8-57 shows the basic connections for DAC control of an amplifier. In this circuit, the DAC is used as the feedback element for an inverting-amplifier configuration. The R-2R ladder digitally adjusts the amount of output signal fed back to the amplifier summing junction. The feedback resistance can be thought of as varying from about 15 k to infinity when the input code changes from full scale to zero. The internal feedback resistor is used as the amplifier input resistor. When the input code is all 0s, the feedback loop is opened and the op-amp output saturates.

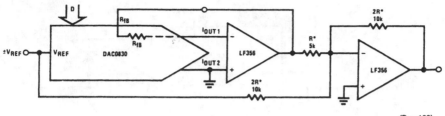

$$V_{OUT} = V_{REF} \frac{(D - 128)}{128}$$

*These resistors are available from Beckman Instruments, Inc. as their part no. 694-3-R10K-Q

$$1 \text{ LSB} = \frac{|V_{REF}|}{128}$$

Input Code		Ideal V_{OUT}	
MSB . . . LSB		$+V_{REF}$	$-V_{REF}$
1 1 1 1 1 1 1 1		$V_{REF} - 1\text{ LSB}$	$-|V_{REF}| + 1\text{ LSB}$
1 1 0 0 0 0 0 0		$V_{REF}/2$	$-|V_{REF}|/2$
1 0 0 0 0 0 0 0		0	0
0 1 1 1 1 1 1 1		-1 LSB	$+1\text{ LSB}$
0 0 1 1 1 1 1 1		$-\dfrac{|V_{REF}|}{2} - 1\text{ LSB}$	$\dfrac{|V_{REF}|}{2} + 1\text{ LSB}$
0 0 0 0 0 0 0 0		$-|V_{REF}|$	$+|V_{REF}|$

FIGURE 8-56 DAC with bipolar output from a fixed reference (National Semiconductor, *Linear Applications Handbook,* 1994, p. 666)

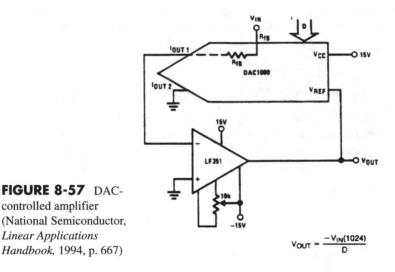

FIGURE 8-57 DAC-controlled amplifier (National Semiconductor, *Linear Applications Handbook,* 1994, p. 667)

$$V_{OUT} = \frac{-V_{IN}(1024)}{D}$$

8.5.6 Capacitance Multiplier

Figure 8-58 shows the basic connections for capacitance multiplication. Actually, the circuit is a DAC-controlled amplifier (used for capacitance multiplication) to give a microprocessor control of system time-domain or frequency-domain response.

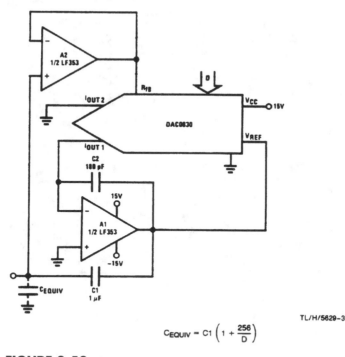

$$C_{EQUIV} = C_1 \left(1 + \frac{256}{D}\right)$$

TL/H/5629-3

FIGURE 8-58 Capacitance multiplier (National Semiconductor, *Linear Applications Handbook,* 1994, p. 667)

In simple terms, the microprocessor controls the digital input to the DAC, which controls the amplifier to produce a variable capacitance. The capacitance can be used to vary the time constant of RC circuits, varying either time or frequency.

In this circuit, the DAC adjusts the gain of a stage with fixed capacitive feedback. This produces a Miller-effect equivalent input capacitance equal to the fixed capacitance multiplied by 1 plus the amplifier gain. The voltage across the equivalent input capacitance to ground is limited to the maximum output voltage of op amp A1, divided by 1 plus $2^n/D$; where n = the DAC bits of resolution, and D = decimal equivalent of the binary input.

8.5.7 High-Voltage Output

Figure 8-59 shows how higher output voltages can be obtained from a DAC (both unipolar and bipolar outputs). The output current of these circuits depends on the current limit of the LM143 op amp (typically 20 mA). Figure 8-60 shows how a discrete power stage can be added to further increase output current capability (to 100 mA at 100 V).

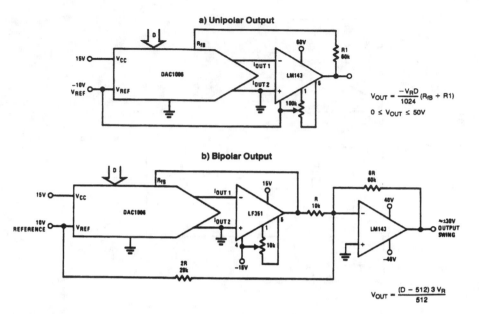

$$V_{OUT} = \frac{-V_R D}{1024}(R_{fB} + R1)$$

$$0 \le V_{OUT} \le 50V$$

$$V_{OUT} = \frac{(D - 512) \, 3 \, V_R}{512}$$

FIGURE 8-59 DAC with high-voltage output (National Semiconductor, *Linear Applications Handbook,* 1994, p. 668)

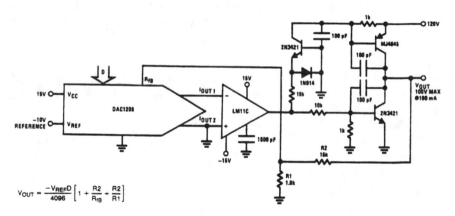

$$V_{OUT} = \frac{-V_{REF} D}{4096}\left[1 + \frac{R2}{R_{fB}} + \frac{R2}{R1}\right]$$

FIGURE 8-60 DAC with high-voltage and increased current output (National Semiconductor, *Linear Applications Handbook,* 1994, p. 668)

8.5.8 High-Current Controller

Figure 8-61 shows a DAC used to provide digital control of a 1-A current sink. Such a circuit can be used for heater control, stepper-motor torque compensation, and automatic test equipment. The largest source of nonlinearity in this circuit is the stability of the current-sensing resistance (with changes in power dissipation). The sens-

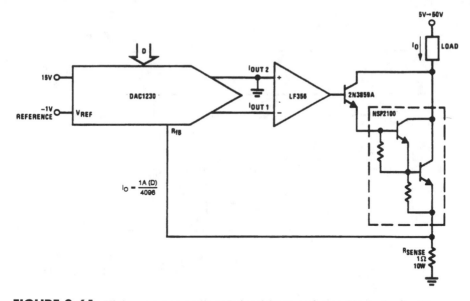

FIGURE 8-61 High-current controller (National Semiconductor, *Linear Applications Handbook,* 1994, p. 669)

ing resistance should be kept as low as possible to minimize this effect. The reference voltage must be reduced (to -1 V, as shown) to maintain the output-current range. The triple Darlington is used to minimize the base-current flowing through the sensing resistance, while simultaneously maintaining the collector current flow to the load.

8.5.9 Current-Loop Controller

Figure 8-62 shows a DAC used to provide digital control of the standard 4-mA to 20-mA industrial-process current loop. The circuit is two-terminal, and all circuit components (including the DAC) are powered directly from the loop.

In this circuit, the output transistor conducts whatever current is necessary to keep the voltage across R3 equal to the voltage across R2. This voltage, and therefore the total loop current, is directly proportional to the output current from the DAC. The net resistance of R1 is used to set the zero-code loop current to 4 mA, and R2 is adjusted to provide the 16-mA output span for a full-scale DAC code. The entire circuit floats by operating at whatever ground-reference potential is required for the total loop resistance and loop current.

To ensure proper operation, the voltage differential between the input and output terminals must be kept in the range of 16 V to 55 V, and the digital inputs to the DAC must be electrically isolated from the ground potential of the controlling microprocessor. This isolation can best be achieved with opto-isolators.

In a non-microprocessor-based system in which the loop-controlling information comes from thumbwheel switches (or a similar mechanical device), the digital

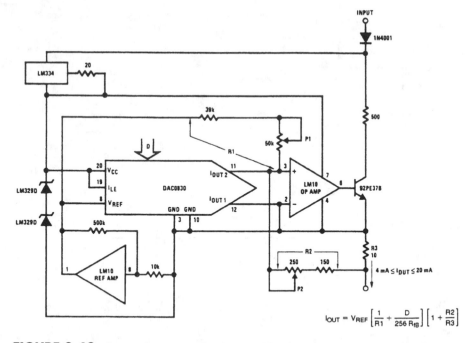

$$I_{OUT} = V_{REF} \left[\frac{1}{R1} + \frac{D}{256\ R_{IB}} \right] \left[1 + \frac{R2}{R3} \right]$$

FIGURE 8-62 Current-loop controller (National Semiconductor, *Linear Applications Handbook,* 1994, p. 670)

input for the DAC can be taken from BCD-to-binary CMOS logic circuits (which are ground referenced to the ground potential of the DAC). The total supply current requirements of all circuits used must (of course) be less than 4 mA, and R1 can be adjusted accordingly.

8.5.10 Tare Compensation

Figure 8-63 shows a DAC (and an ADC) used to provide digital tare compensation. Such a function is used in a weighing system in which the weight of the scale platform, and possibly a container, is subtracted automatically from the total weight being measured. In effect, this expands the range of weight that can be measured by preventing a premature full-scale reading and allows an automatic indication of the actual unknown quantity.

In the basic system of Fig. 8-63, the DAC is initially given a zero code, and the system input is set to a reference quantity. A conversion of the input is performed, then the corresponding code is applied to the DAC. The DAC output then is equal to and of opposite polarity to the input voltage. This forces the amplifier output, and the ADC input, to zero. (In this case, an 8-bit ADC is used.)

The DAC output is held constant so that any subsequent ADC conversion will yield a value relative in magnitude to the initial reference quantity. To ensure that

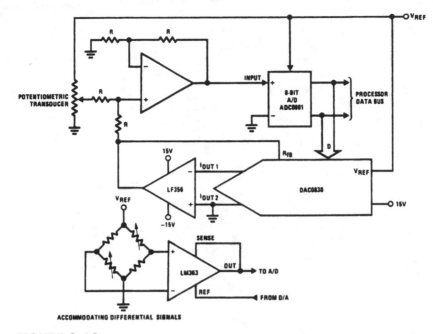

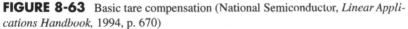

FIGURE 8-63 Basic tare compensation (National Semiconductor, *Linear Applications Handbook,* 1994, p. 670)

the output code from the ADC generates the correct DAC output voltage, the two devices should be driven from the same reference voltage. For differential input signals, an instrumentation amplifier (such as an LM363) can be used. The output reference pin of the LM363 can be driven directly by the DAC as shown. This will offset the ADC input.

CHAPTER 9

Low-Power Data-Acquisition System

This chapter describes an ADC (the MAX192) that contains all major components of a data-acquisition system. Because the ADC requires very low power (1.5-mA operating and 2-μA power-down), the device can provide data acquisition for battery-powered instruments and battery management. The ADC is also well suited for robotic, portable data logging, medical instruments, and any other devices that operate from a single 5-V supply.

9.1 General Description of ADC

Figure 9-1 shows the functional block diagram of the ADC. The IC contains the elements of a 10-bit data-acquisition system that combines an eight-channel multiplexer, high-bandwidth track-and-hold, and serial interface with high conversion speed and ultralow power consumption. The analog inputs are software configurable for single-ended and differential (unipolar-bipolar) operation.

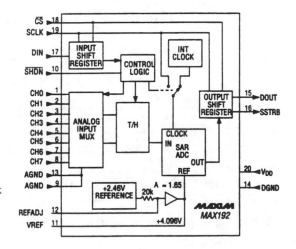

FIGURE 9-1

MAX192 functional block diagram (*Maxim New Releases Data Book,* 1995, p. 7-67)

A four-wire serial interface connects directly to SPI™, QSPI™, and Microwire™ devices by means of external logic. A serial strobe output allows direct connection to TMS320-family digital signal processors. SPI and QSPI are trademarks of Motorola. Microwire is a trademark of National Semiconductor.

The IC uses either an internal clock or an external serial-interface clock to perform SAR ADC conversions. The serial interface can operate beyond 4 MHz when the internal clock is used. The IC has an internal 4.096-V reference with a typical drift of ±30 ppm. A reference buffer-amplifier simplifies gain trim. Two sub-LSBs reduce quantization errors.

A hardwired $\overline{SHDN}$ (shutdown) pin is provided, and there are two software-selectable power-down modes. Accessing the serial interface automatically powers up the device, and the quick turn-on time allows the IC to be shut down between conversions. With powering down between conversions, supply current can be cut to less than 10 μA (at reduced sampling rates).

The device is available in 20-pin DIP and SO packages and in an SSOP that occupies 30% less than an eight-pin DIP.

9.2 Data-Converter Operation

Figure 9-2 shows the pin descriptions for the block diagram of Fig. 9-1. Figure 9-3 shows the IC in a typical operating circuit. As shown, the IC uses a successive-approximation conversion technique and input track-and-hold (T/H) circuitry to con-

PIN	NAME	FUNCTION
1–8	CH0–CH7	Sampling Analog Inputs
9, 13	AGND	Analog Ground. Also IN- Input for single-enabled conversions. Connect both AGND pins to analog ground.
10	$\overline{SHDN}$	Three-Level Shutdown Input. Pulling SHDN low shuts the MAX192 down to 10μA (max) supply current, otherwise the MAX192 is fully operational. Pulling SHDN high puts the reference-buffer amplifier in internal compensation mode. Letting SHDN float puts the reference-buffer amplifier in external compensation mode.
11	VREF	Reference Voltage for analog-to-digital conversion. Also, Output of the Reference Buffer Amplifier. Add a 4.7μF capacitor to ground when using external compensation mode. Also functions as an input when used with a precision external reference.
12	REFADJ	Reference-Buffer Amplifier Input. To disable the reference-buffer amplifier, tie REFADJ to V_{DD}.
14	DGND	Digital Ground
15	DOUT	Serial Data Output. Data is clocked out at the falling edge of SCLK. High impedance when $\overline{CS}$ is high.
16	SSTRB	Serial Strobe Output. In internal clock mode, SSTRB goes low when the MAX192 begins the A/D conversion and goes high when the conversion is done. In external clock mode, SSTRB pulses high for one clock period before the MSB decision. SSTRB is high impedance when $\overline{CS}$ is high (external mode).
17	DIN	Serial Data Input. Data is clocked in at the rising edge of SCLK.
18	$\overline{CS}$	Active-Low Chip Select. Data will not be clocked into DIN unless $\overline{CS}$ is low. When $\overline{CS}$ is high, DOUT is high impedance.
19	SCLK	Serial Clock Input. Clocks data in and out of serial interface. In external clock mode, SCLK also sets the conversion speed. (Duty cycle must be 45% to 55%.)
20	V_{DD}	Positive Supply Voltage, +5V ±5%

FIGURE 9-2 MAX192 pin descriptions (*Maxim New Releases Data Book*, 1995, p. 7-66)

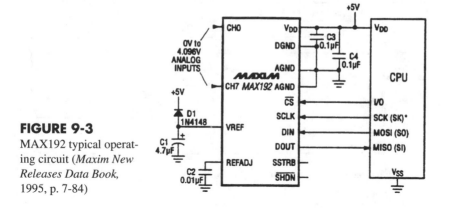

FIGURE 9-3

MAX192 typical operating circuit (*Maxim New Releases Data Book,* 1995, p. 7-84)

vert an analog signal to a 10-bit digital output. A flexible serial interface provides easy interface to microprocessors.

9.2.1 Pseudodifferential Input

Figure 9-4 shows the sampling architecture of the analog input circuit. In the single-ended mode, IN+ is internally switched to CH0-CH7 and IN− is switched to AGND. In the differential mode, IN+ and IN− are selected from pairs of CH0/CH1, CH2/CH3, CH4/CH5, and CH6/CH7. Figures 9-5 and 9-6 show channel selection for single-ended and differential modes, respectively.

In the differential mode, IN− and IN+ are internally switched to either one of the analog inputs. This configuration is pseudodifferential in that only the signal at IN+ is sampled. The return side (IN−) must remain stable within ±0.5 LSB (±0.1 LSB for best results) with respect to AGND during a conversion. To stabilize the return side, connect a 0.1-µF capacitor from AIN− (the selected analog input, respectively) to AGND.

During the acquisition interval, the channel selected as the positive input (IN+) charges capacitor CHOLD. The acquisition interval spans three SCLK cycles

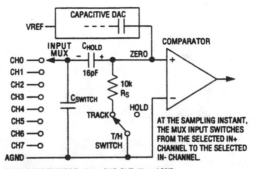

FIGURE 9-4 Sampling architecture of analog input circuit (*Maxim New Releases Data Book,* 1995, p. 7-67)

SINGLE-ENDED MODE: IN+ = CH0-CH7, IN− = AGND.
DIFFERENTIAL MODE (BIPOLAR): IN+ AND IN− SELECTED FROM PAIRS OF CH0/CH1, CH2/CH3, CH4/CH5, CH6/CH7.

SEL2	SEL1	SEL0	CH0	CH1	CH2	CH3	CH4	CH5	CH6	CH7	AGND
0	0	0	+								−
1	0	0		+							−
0	0	1			+						−
1	0	1				+					−
0	1	0					+				−
1	1	0						+			−
0	1	1							+		−
1	1	1								+	−

FIGURE 9-5 Channel selection in single-ended mode (*Maxim New Releases Data Book,* 1995, p. 7-68)

SEL2	SEL1	SEL0	CH0	CH1	CH2	CH3	CH4	CH5	CH6	CH7
0	0	0	+	−						
0	0	1			+	−				
0	1	0					+	−		
0	1	1							+	−
1	0	0	−	+						
1	0	1			−	+				
1	1	0					−	+		
1	1	1							−	+

FIGURE 9-6 Channel selection in differential mode (*Maxim New Releases Data Book,* 1995, p. 7-69)

and ends on the falling clock (SCLK) edge after the last bit of the input control word has been entered. At the end of the acquisition interval, the T/H switch opens, retaining the charge on CHOLD as a sample of the signal at N+.

The conversion interval begins with the input multiplexer switching CHOLD from the positive input (IN+) to the negative input (IN−). In the single-ended mode, IN− is simply at AGND. This unbalances the ZERO node at the comparator input. The capacitive DAC adjusts during the remainder of the conversion cycle to restore the ZERO node to 0 V (within the resolution limits). This action is equivalent to transferring a charge of 16 pF $\times$ ($V_{IN}+$ − $V_{IN}-$) from CHOLD to the binary-weighted capacitive DAC, which forms a digital representation of the analog input signal.

9.2.2 Track and Hold

The T/H function enters the tracking mode on the falling clock edge after the fifth bit of the 8-bit control word has been shifted in. The T/H enters the hold mode on the falling clock edge after the eighth bit of the control word has been shifted in.

If the converter is set up for single-ended inputs, IN− is connected to AGND, and the converter samples the +input. If the converter is set up for differential inputs, IN− connects to the −input, and the difference of IN+ − IN− is sampled. At the end of conversion, the positive input connects back to IN+, and CHOLD charges to the input signal.

The time required for the T/H to acquire an input signal is a function of how quickly the input capacitance is charged. If the input-signal source impedance is high, the acquisition time lengthens, and more time must be allowed between conversions. Acquisition time is calculated as follows:

$$tAZ = 9 \, (RS + RIN) \, 16 \, pF$$

where RIN = 5 k, RS = the source impedance of the input signal, and tAZ is never less than 1.5 μs. Source impedances less than 5 k do not significantly affect the AC performance of the ADC.

9.2.3 Input Bandwidth

The input-tracking circuits have a 4.5-MHz small-signal bandwidth. This makes it possible to use undersampling techniques to digitize high-speed transient events and measure periodic signals with bandwidths that exceed the ADC sampling rate. As discussed in Chapter 4, undersampling can cause aliasing problems, so anti-aliasing filters are recommended. (See the data sheets for the MAX291-MAX297 filters.)

9.2.4 Analog Input Range and Input Protection

Internal protection diodes, which clamp the analog input to V_{DD} and AGND, allow the channel input pins to swing from AGND − 0.3 V to V_{DD} + 0.3 V without damage. However, for accurate conversions near full scale, the inputs must not exceed V_{DD} by more than 50 mV or be lower than AGND by 50 mV.

It is important that if an off-channel analog input exceeds the supplies by more than 50 mV, current will flow through the protection diodes on that input. If this current exceeds 2 mA, the accuracy of the on-channel conversion will be degraded.

The IC can be configured for differential (unipolar or bipolar) or single-ended (unipolar only) inputs, as selected by bits 2 and 3 of the control byte. Figure 9-7 shows the control-byte format. Tables 1 and 2 in Fig. 9-7 refer to Figs. 9-5 and 9-6 in this book.

In the single-ended mode, set the UNI/BIP bit to unipolar. In this mode, analog inputs are internally referenced to AGND with a full-scale input range from 0 V to V_{REF}.

In the differential mode, both unipolar and bipolar settings can be used. Choosing unipolar mode sets the differential input range at 0 V to V_{REF}. The output code is invalid (code 0) when a negative differential input is applied.

The bipolar mode sets the differential input range to ±V_{REF}/2. In the bipolar differential mode, the common-mode input range includes both supply rails. Figure 9-8 shows both unipolar and differential-bipolar ranges.

Bit 7 (MSB)	Bit 6	Bit 5	Bit 4	Bit 3	Bit 2	Bit 1	Bit 0 (LSB)
START	SEL2	SEL1	SEL0	UNI/$\overline{\text{BIP}}$	SGL/$\overline{\text{DIF}}$	PD1	PD0

Bit	Name	Description
7(MSB)	START	The first logic "1" bit after $\overline{\text{CS}}$ goes low defines the beginning of the control byte.
6 5 4	SEL2 SEL1 SEL0	These three bits select which of the eight channels are used for the conversion. See Tables 1 and 2.
3	UNI/$\overline{\text{BIP}}$	1 = unipolar, 0 = bipolar. Selects unipolar or bipolar conversion mode. In unipolar mode, an analog input signal from 0V to VREF can be converted; in differential bipolar mode, the differential signal can range from -VREF / 2 to +VREF / 2. Select differential operation if bipolar mode is used.
2	SGL/$\overline{\text{DIF}}$	1 = single ended, 0 = differential. Selects single-ended or differential conversions. In single-ended mode, input signal voltages are referred to AGND. In differential mode, the voltage difference between two channels is measured. Select unipolar operation if single-ended mode is used. See Tables 1 and 2.
1 0(LSB)	PD1 PD0	Selects clock and power-down modes. PD1 PD0 Mode 0 0 Full power-down ($I_Q = 2\mu A$) 0 1 Fast power-down ($I_Q = 30\mu A$) 1 0 Internal clock mode 1 1 External clock mode

FIGURE 9-7 Control-byte format (*Maxim New Releases Data Book,* 1995, p. 7-69)

9.2.5 A Quick-Look Circuit

Figure 9-9 shows a test circuit for evaluating performance of the IC. The IC requires that a control byte be written to D_{IN} before each conversion. Tying D_{IN} to +5 V feeds in continuous control bytes that trigger single-ended conversions on CH7 (in external clock mode) without powering down between conversions.

REFERENCE		ZERO SCALE	FULL SCALE
Internal Reference		0V	+4.096V
External Reference	at REFADJ	0V	V_{REFADJ} (1.678)
	at VREF	0V	V_{REF}

FIGURE 9-8 Unipolar and differential bipolar ranges (*Maxim New Releases Data Book,* 1995, p. 7-70)

REFERENCE		NEGATIVE FULL SCALE	ZERO SCALE	FULL SCALE
Internal Reference		-4.096V / 2	0V	+4.096V / 2
External Reference	at REFADJ	$-1/2V_{REFADJ}$ (1.678)	0V	$+1/2V_{REFADJ}$ (1.678)
	at VREF	$-1/2\ V_{REF}$	0V	$+1/2\ V_{REF}$

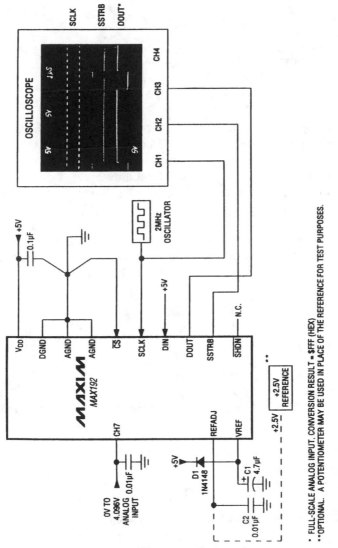

* FULL-SCALE ANALOG INPUT. CONVERSION RESULT = $FFF (HEX)
** OPTIONAL.   A POTENTIOMETER MAY BE USED IN PLACE OF THE REFERENCE FOR TEST PURPOSES.

FIGURE 9-9 Quick-look circuit (*Maxim New Releases Data Book*, 1995, p. 7-71)

In the external-clock mode, the SSTRB output pulses high for one clock period before the MSB of the conversion result comes out of D_{OUT}. Varying the analog input to CH7 should alter the sequence of bits from D_{OUT}. A total of 15 clock cycles is required per conversion. All transitions of the SSTRB and D_{OUT} outputs occur on the falling edge of SCLK.

9.2.6 Starting a Conversion

One starts a conversion by clocking a control byte (Fig. 9-7) into D_{IN}. Each rising edge on SCLK, with $\overline{CS}$ low, clocks a bit from D_{IN} into the internal shift register. After $\overline{CS}$ falls, the first arriving logic-1 bit defines the MSB of the control byte. Until this first "start" bit arrives, any number of logic-0 bits can be clocked into D_{IN} with no effect.

For operation with SPI, select the correct clock polarity and sampling edge in the SPI control registers (set CPOL and CPHA to 0). Microwire and SPI both transmit a byte and receive a byte at the same time. Using Fig. 9-3 as an example, the simplest software interface requires only three 8-bit transfers to perform a conversion (one 8-bit transfer to configure the ADC and two more 8-bit transfers to clock out the 12-bit conversion result).

The following sequence shows how the operating circuit of Fig. 9-3 can be set up to provide a conversion. Figure 9-10 shows timing for the sequence. Make certain that the CPU serial interface runs in a master mode so that the CPU generates the serial clock. Choose a clock frequency from 100 kHz to 2 MHz.

1. Set up the control byte for external clock mode (Fig. 9-7). Call this TB1. TB1 should be of the format: 1XXXXX11 binary, where the Xs denote the particular channel and conversion-mode selected. (The MSB 1 defines the beginning of the control byte; the last two 1s, bit-1 and bit-0, indicate external clock.)
2. Use a general-purpose I/O line on the CPU to pull CS low (to select the MAX192).
3. Transmit TB1 and simultaneously receive a byte. Call this byte RB1 (and ignore it).

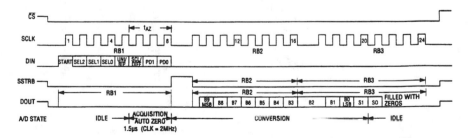

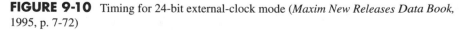

FIGURE 9-10 Timing for 24-bit external-clock mode (*Maxim New Releases Data Book,* 1995, p. 7-72)

4. Transmit a byte of all zeros, and simultaneously receive byte RB2.
5. Transmit a byte of all zeros, and simultaneously receive byte RB3.
6. Pull CS high.

As shown in Fig. 9-10, bytes RB2 and RB3 contain the result of the conversion, padded with one leading 0, two sub-LSB bits, and three trailing 0s. The total conversion time is a function of the serial-clock frequency and the amount of dead time between 8-bit transfers. Make sure that the conversion time does not exceed 120 μs.

Figures 9-11 and 9-12 show the transfer functions for unipolar and differential-bipolar functions, respectively. For a unipolar input, the digital output is straight binary. For bipolar inputs in the differential mode, the digital output is two's-complement. In both cases, data bits are clocked out at the falling edge of SCLK in MSB-first format.

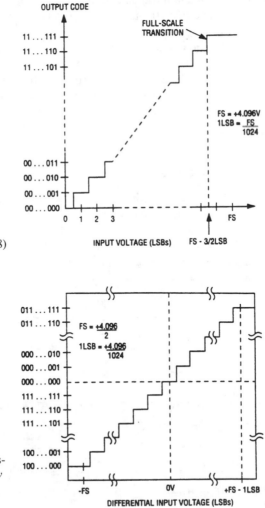

FIGURE 9-11 Unipolar transfer function (*Maxim New Releases Data Book,* 1995, p. 7-78)

FIGURE 9-12 Differential-bipolar transfer function (*Maxim New Releases Data Book,* 1995, p. 7-78)

9.2.7 Clock Modes

The IC can use either an external serial clock or an internal clock to perform conversion. However, in both clock modes, the external clock shifts data in and out of the IC. The T/H acquires the input signal as the last three bits of the control bytes are clocked in to D_{IN}. Bits PD1 and PD0 of the control byte (Fig. 9-7) program the clock mode.

9.2.8 External Clock

Figures 9-13 and 9-14 show the serial-interface and external-clock timing, respectively. In the external-clock mode, the external clock not only shifts data in and out but also drives the ADC conversion process. SSTRB pulses (Fig. 9-2) go high for one clock period after the last bit of the control byte. Successive-approximation bit decisions are made and appear at D_{OUT} on each of the next 12 SCLK falling edges (see Fig. 9-10). The first 10 bits are the true data bits, and the last two are sub-LSB bits.

SSTRB and D_{OUT} go into a high-impedance state when $\overline{CS}$ goes high, after the next $\overline{CS}$ falling edge. SSTRB outputs a logic low.

The conversion must be complete in a minimum time. If not, the droop on the S/H capacitors might degrade conversion results. As a simplified-design guideline, use the internal-clock mode if the clock period exceeds 10 µs, or if the serial-clock interruptions could cause the conversion interval to exceed 120 µs.

9.2.9 Internal Clock

Figures 9-15 and 9-16 show the internal-clock timing and SSTRB (internal-clock) timing, respectively. In the internal-clock mode, the IC uses the internal clock for processing the conversion steps and the external clock for shifting data in and out. This frees the microprocessor from running the SAR-conversion clock and allows conversion results to be read back at the convenience of the processor at any clock

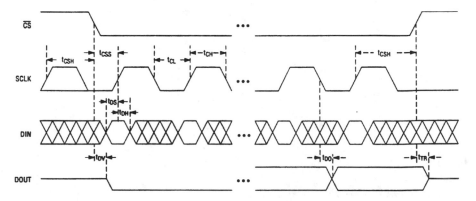

FIGURE 9-13 Serial-interface timing (*Maxim New Releases Data Book,* 1995, p. 7-72)

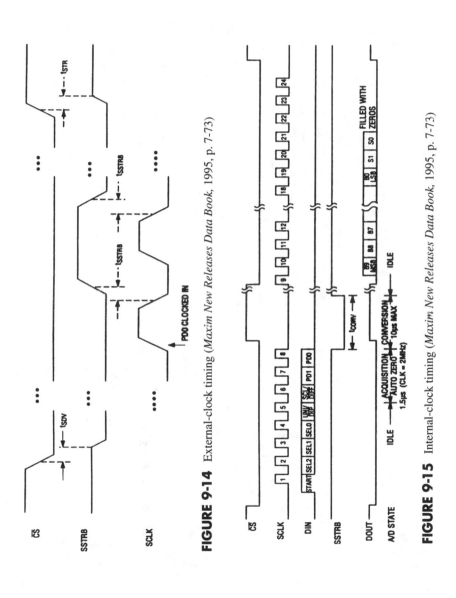

FIGURE 9-14 External-clock timing (*Maxim New Releases Data Book*, 1995, p. 7-73)

FIGURE 9-15 Internal-clock timing (*Maxim New Releases Data Book*, 1995, p. 7-73)

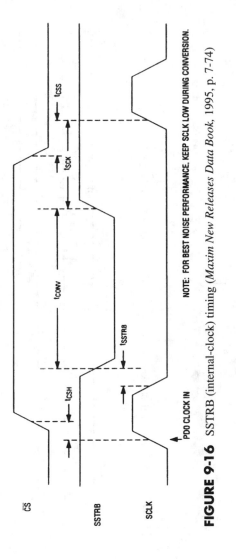

FIGURE 9-16 SSTRB (internal-clock) timing (*Maxim New Releases Data Book*, 1995, p. 7-74)

rate from zero to about 10 MHz.

SSTRB goes low at the start of the conversion and then goes high when the conversion is complete. SSTRB is low for a maximum of 10 μs, during which time SCLK should remain low (for best noise performance). An internal register stores data when the conversion is in progress. SCLK clocks the data out of this register at any time after the conversion is complete.

After SSTRB goes high, the next falling clock edge produces the MSB of the conversion at D_{OUT} followed by the remaining bits in MSB-first format, as shown in Fig. 9-15. $\overline{CS}$ does not have to be held low after a conversion is started.

Pulling $\overline{CS}$ high prevents data from being clocked into the IC but does not adversely affect an internal-clock conversion already in progress. When internal-clock mode is selected, SSTRB does not go into a high-impedance state when $\overline{CS}$ goes high.

In the internal-clock mode, data can be shifted in and out of the IC at clock rates exceeding 4.0 MHz, provided the minimum acquisition time (tAZ) is kept above 1.5 μs (see Section 9.2.2).

9.2.10 Data Framing

Figure 9-17 shows the serial-interface timing necessary to perform a conversion every 15 SCLK cycles in the external-clock mode. Figure 9-18 shows the timing for conversion every 16 SCLK cycles in external clock. As shown, the falling edge of $\overline{CS}$ does not start a conversion. Instead, the first logic-high clocked into D_{IN} is interpreted as a start bit and defines the first bit of the control byte (see Fig. 9-7). A conversion starts on the falling edge of SCLK, after the eighth bit of the control byte (the PD0 bit) is clocked into D_{IN}. The start bit is defined as follows:

The first high bit clocked into D_{IN} with $\overline{CS}$ low any time the converter is idle (after V_{DD} is applied)

or

The first high bit clocked into D_{IN} after bit 3 of a conversion in progress is clocked onto the D_{OUT} pin.

If a falling edge on $\overline{CS}$ forces a start bit before bit 3 (B3) becomes available, then the current conversion is terminated, and a new conversion is started. As a result, the fastest that the MAX192 can run is 15 clocks per conversion. Many microprocessors require that conversions occur in multiplies of eight SCLK clocks. As a result, 16 clocks per conversion typically is the fastest that a microprocessor can drive the MAX192.

9.3 Applications Information

The rest of this chapter is devoted to applications information for the MAX192.

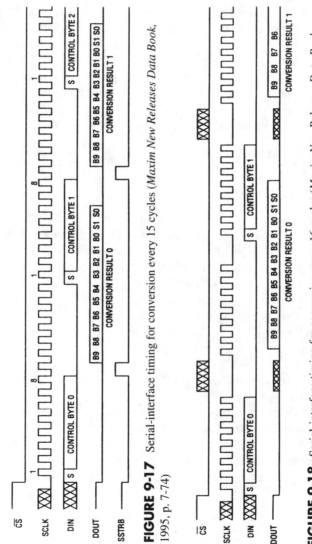

FIGURE 9-17 Serial-interface timing for conversion every 15 cycles (*Maxim New Releases Data Book,* 1995, p. 7-74)

FIGURE 9-18 Serial-interface timing for conversion every 16 cycles (*Maxim New Releases Data Book,* 1995, p. 7-74)

9.3.1 Power-On Reset

When power is first applied, and if $\overline{\text{SHDN}}$ is not pulled low, internal power-on reset circuits activate the MAX192 in the internal-clock mode. The IC is ready to convert when SSTRB goes high. After the power supplies have been stabilized, the internal reset time is 100 μs, and no conversion should be performed during this phase. SSTRB is high on power-up, and if $\overline{\text{CS}}$ is low, the first logic-1 on D_{IN} will be interpreted as a start bit. Until a conversion takes place, D_{OUT} shifts out 0s.

9.3.2 Reference-Buffer Compensation

In addition to the shutdown function, the SHDN pin selects internal or external compensation. This compensation affects both power-up time and maximum conversion speed. Compensated or not, the minimum clock rate is 100 kHz because of the S/H charge-discharge.

To select external compensation, leave $\overline{\text{SHDN}}$ floating as shown in Fig. 9-3. Also use a 4.7-μF (or larger value) capacitor at V_{REF} as shown. This ensures stability and allows operation of the converter at the full clock speed of 2 MHz. External compensation increases power-up time (see Section 9.3.3).

To select internal compensation, pull $\overline{\text{SHDN}}$ high, and do not use an external capacitor at V_{REF}. Internal compensation allows the shortest power-up time but is available only with an external clock and reduces the maximum clock rate to 400 kHz.

9.3.3 Choosing Power-Down Mode

Figure 9-19 shows how the choice of reference-buffer compensation and power-down mode affects both power-up delay and maximum sample rate. Figure 9-20 shows the software codes for selecting shutdown and clock modes. Figure 9-21 shows the code for hard-wired shutdown and compensation mode. Figure 14c referred to in Fig. 9-19 is Fig. 9-22 in this book.

It is possible to save power by placing a converter in a low-current shutdown or power-down state between conversions. In this converter, you can select (through software) either full power-down or fast power-down via bits 7 and 8 of the D_{IN} control byte when $\overline{\text{SHDN}}$ is high, as shown in Figs. 9-7 and 9-20. When $\overline{\text{SHDN}}$ is pulled low, the converter is completely shut down. This is because $\overline{\text{SHDN}}$ overrides bits 7 and 8 of the D_{IN} word, as shown in Fig. 9-21. Full power-down mode turns off all chip functions that draw quiescent current, typically reducing I_{DD} to 2 μA.

Fast power-down mode turns off all circuits except the reference. With fast power-down, the supply current is 30 μA. When internal compensation is used with fast power-down, the power-up time is about 5 μs. Keep in mind that in both software shutdown modes (full or fast), the serial interface remains operational, but the ADC does not convert.

Reference Buffer	Reference-Buffer Compensation Mode	VREF Capacitor (µF)	Power-Down Mode	Power-Up Delay (sec)	Maximum Sampling Rate (ksps)
Enabled	Internal		Fast	5µ	26
Enabled	Internal		Full	300µ	26
Enabled	External	4.7	Fast	See Figure 14c	133
Enabled	External	4.7	Full	See Figure 14c	133
Disabled			Fast	2µ	133
Disabled			Full	2µ	133

FIGURE 9-19 Power-up delay times (*Maxim New Releases Data Book,* 1995, p. 7-76)

FIGURE 9-20 Software codes for shutdown and clock modes (*Maxim New Releases Data Book,* 1995, p. 7-76)

PD1	PD0	Device Mode
1	1	External Clock Mode
1	0	Internal Clock Mode
0	1	Fast Power-Down Mode
0	0	Full Power-Down Mode

FIGURE 9-21 Hardwired shutdown and compensation codes (*Maxim New Releases Data Book,* 1995, p. 7-76)

SHDN State	Device Mode	Reference-Buffer Compensation
1	Enabled	Internal Compensation
Floating	Enabled	External Compensation
0	Full Power-Down	N/A

FIGURE 9-22 Typical power-up delay versus time in shutdown (*Maxim New Releases Data Book,* 1995, p. 7-77)

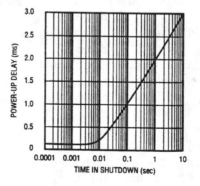

Figures 9-23 and 9-24 show the timing diagrams for power-down using external-clock and internal-clock modes, respectively. Figure 9-25 shows the overall power-up sequence for both full and fast power-down modes (FULLPD and FASTPD).

When external compensation is used (4.7-µF capacitor at V_{REF}; see Fig. 9-3), and the capacitor is fully discharged, the power-up time is 20 ms. In fast power-down,

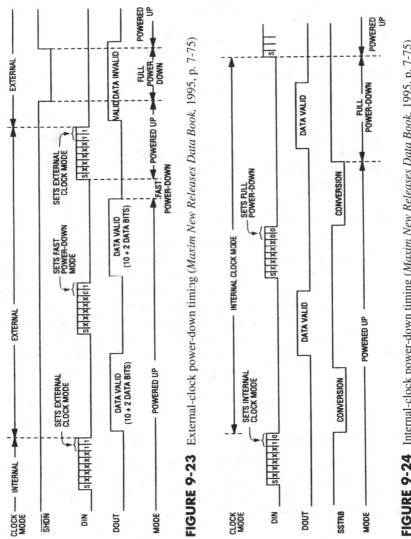

FIGURE 9-23 External-clock power-down timing (*Maxim New Releases Data Book*, 1995, p. 7-75)

FIGURE 9-24 Internal-clock power-down timing (*Maxim New Releases Data Book*, 1995, p. 7-75)

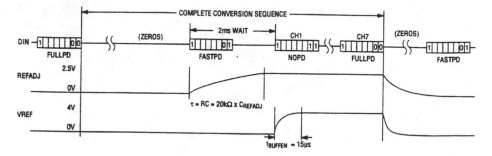

FIGURE 9-25 Overall power-up sequence (*Maxim New Releases Data Book,* 1995, p. 7-77)

the start-up time can be minimized by use of low-leakage capacitors that will not discharge more than ½ LSB during shutdown. (Remember that the capacitor must supply the current into the reference, about 1.5 μA, during shutdown, and must supply transient current at power-up.)

9.3.4 Software Power-Down

As shown in Fig. 9-20, control-byte bits PD0 and PD1 select both the power-down mode and clock mode. When software power-down is used, the ADC continues to operate in the last-specified clock mode until conversion is complete. Then the ADC powers down into a low quiescent state. In the internal-clock mode, the interface remains active, and conversion results can be clocked out during software power-down.

The first logic-1 bit on MD$_{IN}$ is interpreted as a start bit. This powers up the IC. After the start bit, the data-input word, or control byte, also determines clock and power-down modes. For example, if the D$_{IN}$ word is such that PD1 is a logic-1, then the IC remains powered up. If PD1 is a logic-0, a power-down resumes after one conversion.

9.3.5 Hardware Power-Down

As shown in Fig. 9-21, the $\overline{\text{SHDN}}$ pin selects either internal or external reference compensation and places the IC in full power-down mode. Unlike software power-down, conversion is not complete during hardware power-down. Instead, conversion stops immediately when $\overline{\text{SHDN}}$ is brought low.

9.3.6 Power Consumption with Full Power-Down

Figure 9-26 shows the power consumption for one- or eight-channel conversions using full power-down mode and internal reference compensation. A 0.01-μF bypass capacitor (C2 in Fig. 9-3) at REFADJ forms an RC filter with the internal 20-k reference resistor, resulting in a 0.2-ms time constant.

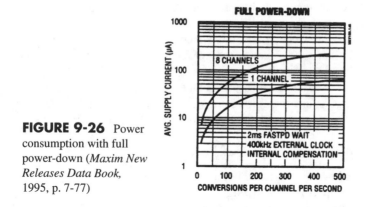

FIGURE 9-26 Power consumption with full power-down (*Maxim New Releases Data Book,* 1995, p. 7-77)

To achieve full 10-bit accuracy, 10 time constants, or 2 ms, are required after power-up. Waiting 2 ms in the fast power-down mode instead of full power-up reduces the power consumption by a factor of 10 or more. This is done by means of the sequence shown in Fig. 9-25.

9.3.7 Power Consumption with Fast Power-Down

Figure 9-27 shows the power consumption with external-reference compensation in fast power-down with one and eight channels converted. The external 4.7-μF compensation (Fig. 9-3) requires a 50-μs wait after power-up (accomplished by 75 idle clocks after a dummy conversion). This configuration combines fast multichannel conversion with the lowest power consumption.

9.3.8 Full Power-Down versus Fast Power-Down

As shown in Figs. 9-26 and 9-27, the tradeoff from a simplified-design standpoint is speed versus power. If power consumption is critical, keep the clock (and conversion speed) less than 500 kHz and use full power-down (Fig. 9-26). This

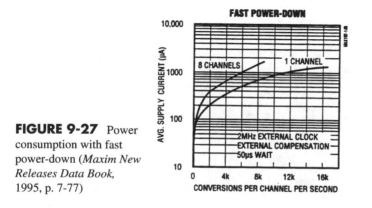

FIGURE 9-27 Power consumption with fast power-down (*Maxim New Releases Data Book,* 1995, p. 7-77)

allows eight channels of conversion (with more than 400 conversions per channel per second), with an average supply current of 200 μA, or less. If speed is all-important, use fast power-down and an external clock of 2 MHz. This allows eight channels of conversion (with more than 8,000 conversions per channel per second) with about 1,500 μA of supply current.

The full power-down mode can be used in other configurations. An example is a situation in which the IC is inactive for long periods of time but intermittent bursts of high-speed conversions are required. Such a change must be made in software.

9.3.9 External versus Internal Reference

Either an external or the internal reference can be used. Diode D1 (see Fig. 9-3) ensures correct startup. Any standard signal diode can be used for D1. When an external reference is required, the reference can be connected directly at the V_{REF} terminal or at the REFADJ pin. (Either pin can be used because the internally trimmed 2.46-V reference is buffered with a gain of 1.678 to scale an external 2.5-V reference at REFADJ to 4.096 V at V_{REF}, as shown in Fig. 9-1.)

With internal reference, the full-scale range is 4.096 V for unipolar inputs and ±2.048 V for differential bipolar inputs. The internal reference can be adjusted using the circuit of Fig. 9-28. This circuit provides for ±1.5% adjustment of the internal reference.

With external reference, the reference is connected to the input of the internal buffer-amplifier (REFADJ) or to the output at V_{REF} (see Fig. 9-1). The REFADJ input impedance is typically 20 k. The V_{REF} impedance is a minimum of 12 k (for DC).

During conversion, an external reference at V_{REF} must be able to deliver up to 350 μA DC load current and have an output impedance of 10 ohms or less. If the reference has higher output impedance or is noisy, bypass the reference close to the V_{REF} pin with a 4.7-μF capacitor (see Fig. 9-3).

It is not necessary to use an external buffer when the reference is applied to the buffered REFADJ input. However, when using the direct V_{REF} input, disable the internal buffer by tying REFADJ to V_{DD}.

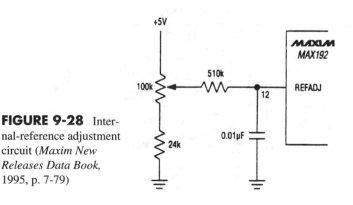

FIGURE 9-28 Internal-reference adjustment circuit (*Maxim New Releases Data Book,* 1995, p. 7-79)

9.3.10 Transfer Function and Gain Adjust

Figure 9-11 shows the nominal, unipolar I/O transfer function in which a typical circuit is used, such as that of Fig. 9-1. Figure 9-12 shows the differential bipolar I/O transfer function. Code transitions occur halfway between successive integer LSB values. The adjustment circuit for the internal reference (see Fig. 9-28) can be used to set the ADC gain within $\pm 1.5\%$. This corresponds to ± 15 LSBs of gain-adjustment range.

9.3.11 Layout, Grounding, and Bypassing

Figure 9-29 shows the recommended grounding and bypassing connections. A single-point analog ground ("star" ground point) should be established at AGND separate from the logic ground. All other analog grounds and the DGND should be connected to this ground. No other digital-system grounds should be connected to the single-point analog ground. The ground return to the power supply for this ground should be low impedance and as short as possible for noise-free operation.

High-frequency noise in the V_{DD} power supply ($+5$ V) can affect the high-speed comparator in the ADC. If the $+5$-V supply is very noisy, connect a 10-ohm resistor as shown in Fig. 9-29. The resistor acts as a low-pass filter. Also bypass the $+5$-V supply to the single-point analog ground with 0.1-μF and 4.7-μF bypass capacitors as shown. Keep the capacitors as close to the IC as possible. The best noise rejection is when the capacitor leads are as short as possible.

As always, for best performance, use PC boards, not wire-wrap boards. Make sure that digital and analog signal lines are separated from each other. Do not run analog and digital (especially clock) lines parallel to one another or digital lines beneath the ADC package.

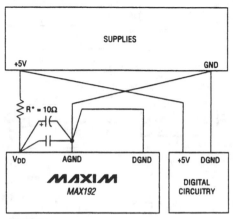

FIGURE 9-29 Recommended grounding and bypassing connections (*Maxim New Releases Data Book,* 1995, p. 7-79)

9.3.12 Interfacing with QSPI

Figure 9-30 shows the interconnections between the MAX192 and a Motorola MC68HC16 for QSPI operation. Figures 9-31 and 9-32 show the assembly code listing. (I have not verified this listing.) Figure 9-33 shows timing for the connections of Fig. 9-30 and the listing of Figs. 9-31 and 9-32.

This QSPI configuration can be programmed to perform a conversion on each of the eight channels. The result is stored in memory without taxing the CPU because QSPI incorporates its own microsequencer.

In the external-clock mode, the MAX192 performs a single-ended, unipolar conversion on each of the eight analog-input channels. The first byte clocked into the MAX192 is the control byte, which triggers the first conversion on CH0. The last two bytes clocked into the MAX192 are all 0. These last two bytes clock out the results of the CH7 conversion.

9.3.13 Interfacing with TMS320

Figure 9-34 shows the basic interconnections between the MAX192 and a TMS320. The circuit is operating in the external-clock mode. Use the following steps to initiate a conversion in the MAX192 and to read the results. Figure 9-35 shows the serial-interface timing.

1. Configure the TMS320 with CLKX (transmit clock) as an active-high output clock and CLKR (TMS320 receive clock) as an active-high input clock. CLKX and CLKR of the TMS320 are tied together with the SCLK input of the MAX192.

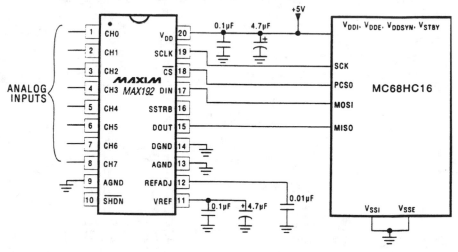

FIGURE 9-30 Interfacing for QSPI operation (*Maxim New Releases Data Book,* 1995, p. 7-80)

```
* Description :
*           This is a shell program for using a stand-alone 68HC16 without any external memory. The internal 1K RAM
*           is put into bank $0F to maintain 68HC11 code compatibility. This program was written with software
*           provided in the Motorola 68HC16 Evaluation Kit.
*
* Roger J.A. Chen, Applications Engineer
* MAXIM Integrated Products
* November 20, 1992
*
•••••••••••••••••••••••••••••••••••••••••••••••••••••••••••••••••••••••••••••••••••••••••••••••••••••
        INCLUDE     'EQUATES.ASM'   ;Equates for common reg addrs
        INCLUDE     'ORG00000.ASM'  ;initialize reset vector
        INCLUDE     'ORG00008.ASM'  ;initialize interrupt vectors
        ORG  $0200                  ;start program after interrupt vectors
        INCLUDE 'INITSYS.ASM'       ;set EK=F,XK=0,YK=0,ZK=0
                                    ;set sys clock at 16.78 MHz, COP off
        INCLUDE 'INITRAM.ASM'       ;turn on internal SRAM at $10000
                                    ;set stack (SK=1, SP=03FE)
MAIN:
        JSR   INITQSPI
MAINLOOP:
        JSR   READ192
WAIT:
        LDAA   SPSR
        ANDA   #$80
        BEQ    WAIT                 ;wait for QSPI to finish
        BRA MAINLOOP
ENDPROGRAM:

INITQSPI:

;This routine sets up the QSPI microsequencer to operate on its own.
;The sequencer will read all eight channels of a MAX192 each time
;it is triggered. The A/D converter results will be left in the
;receive data RAM. Each 16 bit receive data RAM location will
;have a leading zero, 10 + 2 bits of conversion result and three zeros.
;
;Receive RAM Bits 15 14 13 12 11 10 09 08 07 06 05 04 03 02 01 00
;A/D Result         0  MSB                          LSB 0  0  0
***** Initialize the QSPI Registers  ******
        PSHA
        PSHB
        LDAA   #%01111000
        STAA   QPDR                 ;idle state for PCS0-3 = high
        LDAA   #%01111011
        STAA   QPAR                 ;assign port D to be QSPI
        LDAA   #%01111110
        STAA   QDDR                 ;only MISO is an input
        LDD    #$8008
        STD    SPCR0                ;master mode,16 bits/transfer,
                                    ;CPOL=CPHA=0,1MHz Ser Clock
        LDD    #$0000
        STD    SPCR1                ;set delay between PCS0 and SCK,
                                    ;set delay between transfers
```

FIGURE 9-31 Assembly code listing for QSPI operation (*Maxim New Releases Data Book,* 1995, p. 7-81)

2. Drive the MAX192 $\overline{CS}$ low via the XF port of the TMS320 to enable data to be clocked into the D_{IN} of the MAX192.
3. Write an 8-bit word (1XXXXX11) to the MAX192 to initiate a conversion and to place the ADC in the external-clock mode. Refer to Fig. 9-7 to select the proper XXXXX bit values for the specific application.

```
         LDD    #$0800
         STD    SPCR2           ;set ENDQP to $8 for 9 transfers
*****  Initialize QSPI Command RAM  *****

         LDAA   #$80     ;CONT=1,BITSE=0,DT=0,DSCK=0,PCS0=ACTIVE
         STAA   $FD40    ;store first byte in COMMAND RAM
         LDAA   #$C0     ;CONT=1,BITSE=1,DT=0,DSCK=0,PCS0=ACTIVE
         STAA   $FD41
         STAA   $FD42
         STAA   $FD43
         STAA   $FD44
         STAA   $FD45
         STAA   $FD46
         STAA   $FD47
         LDAA   #$40     ;CONT=0,BITSE=1,DT=0,DSCK=0,PCS0=ACTIVE
         STAA   $FD48
*****  Initialize QSPI Transmit RAM *****

         LDD    #$008F
                          STD    $FD20
         LDD    #$00CF
                          STD    $FD22
         LDD    #$009F
                          STD    $FD24
         LDD    #$00DF
                          STD    $FD26
         LDD    #$00AF
                          STD    $FD28
         LDD    #$00EF
                          STD    $FD2A
         LDD    #$00BF
                          STD    $FD2C
         LDD    #$00FF
                          STD    $FD2E
         LDD    #$0000
                          STD    $FD30
         PULB
         PULA
         RTS

READ192:
;This routine triggers the QSPI microsequencer to autonomously
;trigger conversions on all 8 channels of the MAX192. Each
;conversion result is stored in the receive data RAM.
         PSHA
         LDAA   #$80
         ORAA   SPCR1
         STAA   SPCR1      ;just set SPE
         PULA
         RTS

*****  Interrupts/Exceptions  *****

BDM: BGND          ;exception vectors point here
                   ;and put the user in background debug mode
```

FIGURE 9-32 Continued assembly code listing for QSPI operation (*Maxim New Releases Data Book,* 1995, p. 7-82)

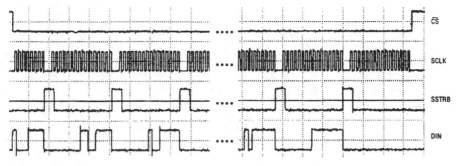

FIGURE 9-33 Timing for QSPI operation (*Maxim New Releases Data Book,* 1995, p. 7-83)

4. Monitor the SSTRB output of the MAX192 through the FSR input of the TMS320. A falling edge on the SSTRB output indicates that the conversion is in progress and data bits are ready to be received from the MAX192. The TMS320 reads in one data bit on each of the next 16 rising edges of SCLK. These data bits represent the 10-bit conversion result and two sub-LSBs, followed by four trailing bits, which should be ignored.
5. Pull $\overline{CS}$ high to disable the MAX192 until the next conversion is initiated.

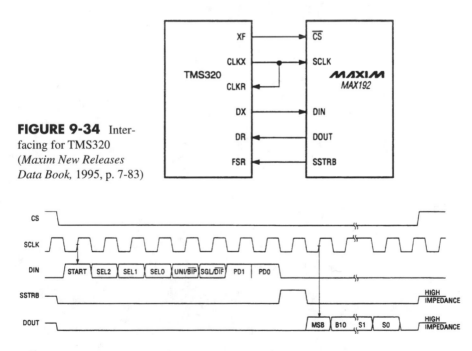

FIGURE 9-34 Interfacing for TMS320 (*Maxim New Releases Data Book,* 1995, p. 7-83)

FIGURE 9-35 Serial-interface timing for TMS320 (*Maxim New Releases Data Book,* 1995, p. 7-83)

CHAPTER 10

Single-Chip Digital Multimeter

This chapter describes an integrating ADC (the MAX133/134) that contains the major components of a 3 ¾-digit digital multimeter (DMM) in a single IC. The ADC can also be used for data-acquisition systems such as data loggers and weigh scales. The internal resolution is ±40,000 counts. An extra digit is supplied as a guard digit to allow autozero or tare of a 4,000-count displayed reading to ¹⁄₁₀ of a displayed count. The conversion time is 50 ms.

The MAX133 and MAX134 differ only in their microprocessor interface. The MAX133 has a 4-bit multiplexed address-data bus, whereas the MAX134 has 3 separate address lines and a 4-bit bidirectional data bus. Both devices can be used with 4-, 8-, and 16-bit microprocessors.

When controlled by a microprocessor, the IC can perform autoranging measurements from ±400.0 mV to ±4000 V full scale. External attenuator resistors are required, but range switching is performed by the ADC. The power supply is typically a 9-V battery, or ±5 V. Operating current is typically 100 µA, with a 25-µA standby current.

10.1 System Considerations

Figure 10-1 shows the typical operating circuit and pin configuration. Figure 10-2 shows a test circuit and basic DMM voltage and current modes. Figure 10-3 shows how the execution of several typical functions is coordinated between the MAX133/134 and the microprocessor.

The MAX133/134 contains an ADC and auxiliary circuitry such as attenuator range switches, a piezoelectric beeper driver, an active filter, a low-battery detector, and both analog and digital power supplies, but it does not include any display-drive capability. The IC reduces component count and system cost by minimizing the external components required for the analog portion of the system, but it does not restrict final production features by including autoranging or other digital-control functions.

The IC is intended to work as the analog front end of a microprocessor; the features of the end product are determined by microprocessor software. The IC provides

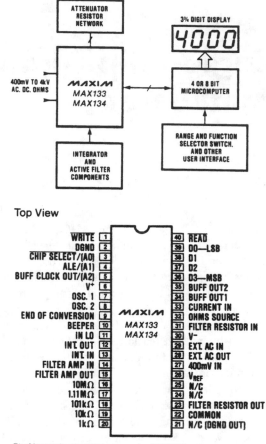

FIGURE 10-1 Typical operating circuit and pin configuration (*Maxim Evaluation Kit Data Book,* 1994, p. 4-79)

Pin Names in parentheses are for MAX134 only.

all the logic and counters for control of the conversion sequence, and the microprocessor does not perform any critical timing of complex control of the IC.

The IC has three range switches for a five-decade attenuator (which uses external resistors) and has additional mode-selection circuits for performing voltage, current, AC or DC, ohms, and continuity measurements. The five-decade attenuator and mode-selection circuits are controlled by the microprocessor through control bits written into the IC.

Figure 10-4 shows the basic elements of the input section. This includes the A/D (analog-to-digital converter), input range switching, and several control circuits.

The IC has a typical mode rejection of the line frequency of at least 80 dB on the voltage ranges; the microprocessor selects rejection of either 50 Hz or 60 Hz by setting a control bit on the IC. A two-pole active filter can also be turned on by the microprocessor, adding about 40 dB normal mode rejection above 50 Hz.

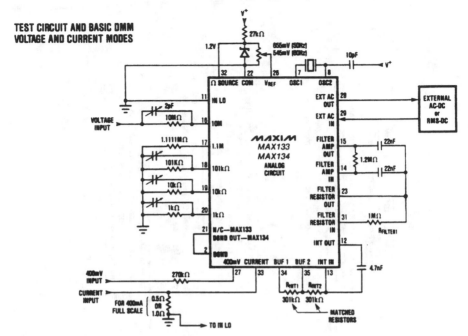

FIGURE 10-2 Test circuit and basic DMM modes (*Maxim Evaluation Kit Data Book,* 1994, p. 4-81)

The basic blocks of the MAX133/134 are as follows:

A/D section
Input range switching
Ohms circuitry
Active filter
Power supply, common, low-battery detector
Oscillator and beeper driver
Digital interface

Each of these blocks or sections is discussed in the following paragraphs.

10.1.1 A/D Section

The A/D (or ADC) uses a "residue-multiplication" conversion scheme to provide a full $\pm 40,000$-count-resolution reading every 50 ms. This is done with the good noise performance and power-line normal-mode rejection associated with integrating ADCs (see Section 10.4 for a description of the conversion method).

FUNCTION	MAX133/134 ACTION	MICROPROCESSOR ACTION
Autoranging	Contains the attenuator control switches. Selects 400mV to 4000V ranges as directed by the microprocessor.	Detects overload and commands the MAX133/134 to select the next higher range. Range switching hysteresis and manual range selection is controlled by the microprocessor.
Zero Reading (system offset correction)	Internally shorts the A/D inputs and performs a measurement of system offset when directed by the microprocessor.	Periodically commands the MAX133/134 to perform a zero reading. Subtracts this zero reading from normal readings to correct for the internal offset of the MAX133/134.
Range/Function Selection	Selects Ohms/Current/ AC-DC/Voltage/Continuity as directed by the microprocessor.	Maintains the user interface, and directs the MAX133/134 to select the desired range.
Display of Readings	Max 133/134 provides raw, non-zero-corrected data to microprocessor.	The microprocessor performs zero correction and any gain correction or scaling that is desired. The microprocessor then displays the information, using either its own display driver capability or an external display driver.
Value added DMM features such as display hold, peak hold, either manual range selection or autoranging, peak reading hold, min/max display, thermocouple linearization, etc.	Performs conversions as directed by the microprocessor, returning the A/D results to the microprocessor.	Uses the MAX133/134 conversion results and software routines to provide a multitude of product features.
Digital panel meter features such as zero and span adjustment, high/low limit alarms, display in engineering units, etc.	Performs conversions and range selection as directed by the microprocessor.	Takes the MAX133/134 readings, performs zero offset and scale corrections, then displays the results. The microprocessor also performs such functions as high/low limit alarms.

FIGURE 10-3 Coordination of MAX133/134 and microprocessor (*Maxim Evaluation Kit Data Book*, 1994, p. 4-82)

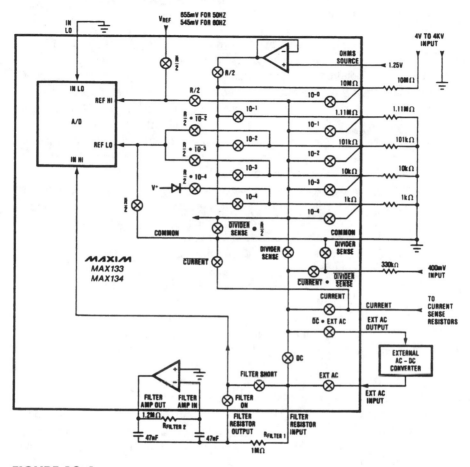

FIGURE 10-4 Basic elements of input section (*Maxim Evaluation Kit Data Book,* 1994, p. 4-84)

All timing and ADC-conversion phase control is performed by the IC without microprocessor intervention. The A/D section performs a non-zero-corrected conversion every 50 ms (20 conversions per second).

The microprocessor must periodically direct the IC to perform a read-zero conversion, which also takes 50 ms. This read-zero conversion is performed with IN LO (low input) internally shorted to IN HI (high input), and the result of zero conversion must be subtracted (by the microprocessor) from normal measurements to provide a zero-corrected reading.

The zero correction that must be subtracted is determined by the MAX133/134 internal offsets. Because these offsets are relatively slow changing, zero conversion readings need only be taken often enough to track long-term drifts and temperature changes. The zero-conversion reading changes slightly with a change in common-mode input voltage or reference voltage, and a new zero-conversion reading should be taken if either of these changes.

In ratiometric ohms measurement, the reference voltage changes considerably as the value of the unknown resistor varies. To reduce the errors caused by the system offset, the IC "chops" the input buffer and integrator. This "chop" consists of a reversal of the input transistors during the conversion cycle. The timing of the chop is such that in the R/2 or ohms-measurement mode, the system offset is almost completely nulled out if the X2 mode is not selected. Even if the X2 mode is selected, the system offset does not exceed 5,000 counts on any range. Because the internal full-scale range is greater than $\pm 49,000$ counts, at least $\pm 40,000$ counts of resolution are available after zero-offset correction.

Each conversion result is latched into a conversion register that can be read by the microprocessor. The data format is nines-complement BCD (a zero reading is 00000, a -1 reading is 99999, a -25000 reading is 75000). The nines-complement form is the most convenient BCD format because the addition of the nines-complement of a number is equivalent to subtracting that number (see Section 10.6 for BCD-to-binary conversion).

The last digit of conversion is used for digital autozero and is usually not displayed. Each count of the least-significant digit of the IC output corresponds to $\frac{1}{10}$ of a count, if a 4,000-count full-scale display is used.

For current ranges with a voltage drop of only 200 mV, the measured reading can be multiplied by two, using the X2 function. The X2 function reduces the RINT resistor value by a factor of two during the integrate phase. With the X2 range, a 200-mV input results in a full-scale 4.000.0 measured reading.

As an alternative, the normal 400-mV range can be used, with the multiplication by two being done (digitally) by the microprocessor. In this case, each count of the least significant digit is $\frac{1}{5}$ of the displayed count. A 100-mV full-scale voltage drop can be achieved using both the X2 range and digital times-two multiplication in the microprocessor.

Each of the 20 conversions per second has a zero-integrator phase to ensure rapid recovery from overload. The IC recovers (to within two counts) one conversion time after an overload of 10 times full scale (when the onboard active filter is not used, see Section 10.1.4).

10.1.2 Input Range Switching

In voltage-measurement ranges other than 400 mV, voltages are applied to the 10-M pin through an external 10-M resistor as shown in Fig. 10-2. With selection of the proper shunt resistance (1.1 M through 1 k), the input voltage is attenuated to 400-mV. The input-attenuator switch section includes analog switches to switch the input current and to sense the voltage on the shunt resistor. Other input-switching functions select between the input-attenuator output and the voltage developed across the current-sensing resistors during measurement.

The 5-pA input-bias current might result in unacceptable errors with a 10-M input resistor on the 400-mV scale, so a separate pin with a 100-k to 1-M input resistor is used. The 10-M resistor used on the higher voltage ranges does not cause appreciable error because the input-leakage current is shunted to ground through the 1.11-M to 1-k attenuator shunt resistors.

To avoid errors that might occur through high-frequency coupling, high-voltage signals from the attenuator input to the 400-mV and current inputs have 10-k switches which connect the inputs to ground when not selected. The input section also includes switches for an external AC-DC converter to be inserted into the signal path (see Section 10.5.3).

10.1.3 Ohms and Diode Measurement

Figure 10-5 shows the basic connections for ohms (resistance) measurement and diode test. The input-attenuator resistors are also used in the ohms mode. (The 10-M resistor must be externally paralleled with other resistors to obtain exactly 1 M, 100 k, etc.) The ohms buffer input (OHMS SOURCE, see Fig. 10-4) is usually connected directly to an external reference or to another 1.25-V source.

In the 4-k through 40-M ranges, there is a total of 1.25-V across the series combination of reference resistor, unknown resistor, and the input-protection network. The maximum voltage across the unknown resistor at full-scale is less than 400 mV. On the 400-ohm range, the ohms voltage source is a diode connected to V+ through a 2-k p-channel switch. With a 3-V common voltage, this supplies about 2.2 V across the series combination of reference resistor, unknown resistor, and input-protection network. This higher voltage is used on the 400-ohm range to compensate for the decrease in reference voltage caused by the input-protection network. The IC operates with PTC (positive temperature coefficient) protection resistors of 2 k or less.

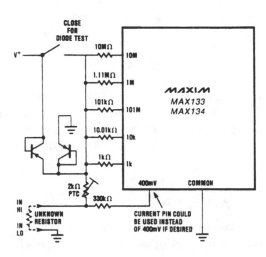

FIGURE 10-5 Basic connections for ohms and diode test (*Maxim Evaluation Kit Data Book,* 1994, p. 4-85)

The voltage across the reference resistor is used as the reference voltage for the A/D in the ohms mode. The differential voltage between IN LO and IN HI is the input signal. The integration period is 500 counts independent of the 50/60-Hz control-bit setting.

The digital output code is as follows:

$$50,000 \times RUNKNOWN/RREF$$

with a maximum non-zero-corrected output code of $\pm 49,500$ and a maximum zero reading of 5,000.

A 1-k reference resistor is used for the 400-ohm full scale, a 10-k reference for a 4-k full scale, and so on. A 10-M reference resistor is used for both the 4-M full scale and the 40-M full scale. To obtain the correct results in the ohms-measurement or R/2 mode, the conversion result must be multiplied by two. This can be done digitally by the microprocessor or by the X2 range except on the 40-M scale. The 40-M range has the same reference resistor as the 4-M range, but a times-10 scale factor is obtained by not multiplying by 2 and by activating the divide-by-5 function.

If the times-five multiplication is performed by the microprocessor, the read-zero offset of the IC in the ohms mode is just a few counts and is nearly independent of the unknown-resistor value. If the X2 mode is used to multiply by 2, then frequent read-zero readings should be taken, because the read-zero offset is inversely proportional to the reference voltage (which varies as the unknown resistor varies).

Because the input-protection resistor shown in Fig. 10-5 reduces the reference and input voltage, particularly on the 400-ohm scale, the PTC resistance should be as low as possible but still must maintain the desired level of protection. A PTC resistance in excess of 2 k increases the noise level for measurements on the 400-ohm range.

Because the IC does not use a reference capacitor, the only limit on the response time in the ohms mode is the active filter. Even when the active filter is turned off, RFILTER1 (see Fig. 10-4) is still connected, and the input voltage must charge the filter capacitors. This generally is noticed only on the 4-M and 40-M ranges.

A diode-test range can be implemented by connecting the PTC (used for input protection in the ohms ranges, see Fig. 10-5) to V+. The PTC then delivers about 1 mA of current to the diode (connected between IN HI and IN LO). The diode voltage can be measured on the standard 4-V scale, or on the 400-mV scale with the divide-by-five function activated to result in a 2-V full scale. As always, the latched-continuity circuit is active, and latches whenever the input voltage goes below about 100 mV. The microprocessor can also test the measured voltage at the end of each conversion, if a more precise detection of continuity threshold is desired.

10.1.4 Active Filter

Figure 10-6 shows the basic connections for the active filter. (Figure 10-4 shows connections to the input section of this two-pole filter.) The op-amp offset has

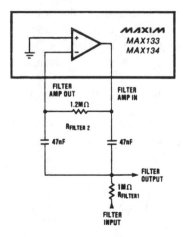

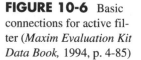

FIGURE 10-6 Basic connections for active filter (*Maxim Evaluation Kit Data Book,* 1994, p. 4-85)

no effect on the DC accuracy because the op amp is AC coupled (only), and the DC signal path is through the passive 1-M resistor.

The active filter limits the response time or speed to input-voltage changes. For that reason, it might be desirable to disconnect the input voltage during autoranging. Because the source impedance at the filter input varies with the input attenuator selected, the response time is slower on the 4-V range.

10.1.5 Oscillator and Beeper Driver

The IC is designed to operate with a 32768-Hz tuning-fork crystal similar to the Statek CX-IV; only one external capacitor and no external resistors are used. If desired, the OSC1 pin can be driven externally.

The 32-kHz clock is used internally as the clock for the sequence and measurement counters. The clock is also divided down to 2,048 Hz and 4,096 Hz to drive a beeper. The beeper output swings from $V+$ to $V-$, and can directly drive piezo-electric beepers.

Two control bits set by the microprocessor select the frequency (2,048 or 4,096 Hz) of the beeper and turn on and turn off the beeper. The beeper is controlled by the microprocessor and can be used for both continuity indication and an audible operator-feedback signal to indicate peak-hold or range changes.

10.1.6 Power Supply and Low-Battery Detector

Although the ICs are designed to operate with a 9-V battery, ± 5-V supplies also can be used. The maximum supply current is 250 μA, with a typical operating current of 100 μA. An on-board low-battery detector indicates when the battery voltage is approaching the minimum operating voltage of about 6.8 V.

As shown in Fig. 10-2, analog common (COM) is taken from a zener and is nominally 3.0 V less than $V+$. For low-cost applications, the common voltage (with a temperature coefficient, or tempco, of 80 ppm/°C) can be used as a reference.

However, in most applications, a bandgap reference is connected to common, with a pullup resistor to V+. A voltage divider connected across the bandgap reference generates the 545-mV (60-Hz operation) or 655-mV (50-Hz operation) reference voltage applied to V_{REF}. In a battery-powered meter, the COM pin is used as the system ground-reference point.

The IC can also generate a digital-ground voltage, which is nominally 5-V below V+. This voltage remains in the range of 5 V ±10%, with a sink capability of 5 µA to 500 µA. The DGND (digital ground) generator has substantial current-sinking capability but can easily be pulled to a more negative voltage because the current-sourcing capability is only 1 µA typical.

The MAX133 internally connects the digital-ground generator to the DGND pin (see Fig. 10-2). Normally, the MAX133 is powered by a 9-V battery, and the ground, V−, or VSS pin of the microprocessor is connected to the MAX133 DGND pin.

The MAX134 connects the DGND voltage generator to the DGND OUT pin. (The MAX134 DGND pin is an input only.) When the MAX134 is used with a 9-V battery, connect the DGND OUT pin to the DGND pin. For use with ± 5-V supplies, connect the DGND pin to ground, V+ to +5 V, and V− to −5 V.

10.2 Digital Interface

The MAX133 and MAX134 differ only in their digital interface. The MAX133 has a multiplexed address and bidirectional data bus. The MAX134 has three separate address lines in addition to a bidirectional data bus. In both ICs, the data bus has 4 bits. This permits use of both ICs with 4-bit or 8-bit microprocessors.

10.2.1 MAX134 Digital Interface

Figure 10-7 shows the read-write sequence between the MAX134 and the microprocessor. A 4-bit bidirectional bus, D0-D3, is required for interface. In addition to the four data-bus lines, there are three address lines (A0-A2), and two control signals—WR (write) and RD (read).

The A0-A2 address lines select one of five control registers. When WR goes low, data bits are written from the bus into the MAX134 control register, addressed by A0-A2. When RD is low, the MAX134 drives the bidirectional bus. The data bits contained in the results or status register (and addressed by A0-A2) are placed on the bus.

10.2.2 MAX133 Digital Interface

Figure 10-8 shows the read/write sequence between the MAX133 and the microprocessor. Only seven lines are required for this interface. The microprocessor first selects the register to be read or written to by placing the register address onto a 4-bit multiplexed address-data bus. The microprocessor then pulses the address latch enable (ALE) line high to latch the register address into the MAX133.

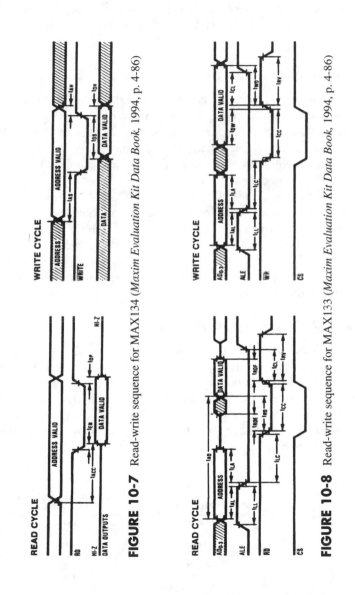

FIGURE 10-7 Read-write sequence for MAX134 (*Maxim Evaluation Kit Data Book*, 1994, p. 4-86)

FIGURE 10-8 Read-write sequence for MAX133 (*Maxim Evaluation Kit Data Book*, 1994, p. 4-86)

To read the selected register, the microprocessor drives the read line low, and the MAX133 places the register data onto the data bus. To write to the selected register, the address is latched as described, then the microprocessor places the data onto the bus. When this is complete, the microprocessor pulls the write line low. The MAX133 latches the data into the selected register on the rising edge of write. The chip-select (CS) line must be low to enable either the RD or WR lines, but ALE is not gated by CS.

10.2.3 MAX133/134 Digital Interface

In most cases, the EOC (end of conversion) signal is either monitored by an I/O pin or drives an interrupt to the microprocessor. In battery-powered systems, it might be desirable to put the microprocessor in a standby mode until EOC goes high. The microprocessor can then perform any required data-processing and display updates and then return to standby. This conserves battery power because the microprocessor power consumption is minimized.

The data bits that have been latched in the MAX133/134 control registers do not immediately affect operation. The input registers are double buffered, and the control bits take effect during the 21st clock cycle after EOC goes high. In the hold mode, the double-buffered registers are transparent, and any updates to the registers take effect immediately, as do any changes made during the one-clock-cycle period at the end of each conversion. (This is the time during which the second rank of buffers are being updated.)

10.2.4 Description of Output Bits

Figure 10-9 shows examples of the nines-complement BCD format used for data output. In addition to data, the latched-continuity output bit is high if the input voltage goes below the continuity threshold of about 100 mV since the last time the register was read. Each time this register (register 5) is read, the continuity latch is reset. The low-battery output bit is high whenever the battery voltage is less than the low-battery detect voltage (about 6.8 V). The holding output bit is low whenever the MAX133/134 is in the hold state.

FIGURE 10-9 Examples of nines-complement BCD format (*Maxim Evaluation Kit Data Book,* 1994, p. 4-87)

MEASUREMENT RESULT	BCD DATA
+40000	40000
—	—
+00100	00100
—	—
+00001	00001
+00000	00000
(there is NO -00000)	
-00001	99999
—	—
-00100	99900
—	—
-40000	60000

10.2.5 Description of Control Bits

Figure 10-10 shows the register map of output data from the MAX133/134 to the microprocessor. Figure 10-11 shows the same information for input data from the microprocessor. The following is a summary of the bit functions.

Hold. A 1 in hold stops conversions at the end of the next conversion. If the MAX133/134 is in the hold mode, a conversion starts on the next clock cycle after hold is set to 0. The oscillator continues to run and all circuitry is active during the hold mode.

High Frequency. A 1 in the high-frequency bit selects 4,096 Hz as the beeper frequency. A 0 selects 2,048 Hz.

Beeper On. A 1 turns the beeper driver on.

Sleep. A 1 in sleep puts the MAX133/134 into the sleep or standby mode. The common voltage buffer is turned off, as is the internal analog circuit, but the DGND circuit is still active. The oscillator continues to run. Current consumption is reduced to 25 μA. Several conversions must be performed after one exits the sleep mode before full conversion accuracy is obtained.

10-0 through 10-4. These bits control the attenuator-network switches. The 10-0 bit selects the 10-M input without activating any shunt resistors. This is an alternative 400-mV input. The 10-1 bit activates the 10:1 attenuation by selecting the 10-M input and connecting the 1.111-M shunt. Similarly, 10-2, 10-3, and 10-4 bits select input attenuation factors of 100, 1,000, and 10,000, respectively. In the ohms mode, these bits set the resistance range.

50Hz. When set to 1, the integration period for voltage measurements is one cycle of the 50-Hz power mains (655 clock cycles). When 0, the integration period is one 60-Hz power-line cycle (545 clock cycles).

X2. Setting the bit to 1 activates the "times-2" function. When X2 is active, RINT2 only is used as the integrator resistor during the integration phase. RINT1 and RINT2 in series are used as the integration phase when X2 is 0. If RINT1 = RINT2, setting the X2 bit doubles the digital output for a given input voltage.

÷ 5. When this bit is set to 1, the integration period is reduced by a factor of 5. This reduces the digital-output code by a factor of 5 and allows a higher input voltage to be used. The full-scale input voltage is multiplied by 5 when this bit is set. Caution should be used to make sure that the 2-μA maximum recommended integrator output current is not exceeded, resulting in degraded linearity.

Ohms or R/2. Setting this bit to 1 selects the ohms measurement mode (see Section 10.1.3). Set the divider sense to 0 for ohms measurement.

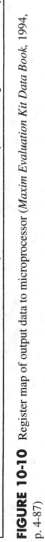

ADDRESS OR REGISTER NUMBER	REGISTER NAME	REGISTER CONTENTS			
0	Ones	Conversion Result			
1	Tens	BCD data for least significant digit (The undisplayed digit used for digital autozero)			
		BCD data of Conversion Result (Least significant displayed digit)			
2	Hundreds	BCD Data of Conversion Result			
3	Thousands	BCD Data of Conversion Result			
4	10 Thousands	BCD Data of Conversion Result			
5	Status	D3	D2	D1	D0
		Always 1	Latched Continuity	$\overline{\text{Holding}}$	Low Battery

FIGURE 10-10 Register map of output data to microprocessor (*Maxim Evaluation Kit Data Book*, 1994, p. 4-87)

ADDRESS OR REGISTER NUMBER	D3	D2	D1	D0
0	Hold	High Frequency	Beeper ON	Sleep
1	10-0	Filter Short	+5	50Hz
2	10-4	10-3	10-2	10-1
3	DC	Ext AC	Divider Sense	Ohms R/2
4	Current	X2	Read Zero	Filter On

FILTER ON	FILTER SHORT	FUNCTION
1	0	Normal filter on condition
1	1	Filter on, $R_{FILTER1}$ is bypassed. Use this bit combination to compensate for the higher source impedance of the 4V range.
0	1	Bypasses the Filter.
0	0	Invalid combination, do not use.

BIT SET	VOLTAGE RANGE	OHMS RANGE
10-0	400mV	4MΩ and 40MΩ
10-1	4V	400kΩ
10-2	40V	40kΩ
10-3	400V	4kΩ
10-4	4000V	400Ω

NOTE: The divider sense bit must also be set to enable the 10-0 through 10-4 bits.

FIGURE 10-11 Register map of input data from microprocessor (*Maxim Evaluation Kit Data Book*, 1994, p. 4-88)

Read Zero. Setting this bit to 1 causes the next conversion to be a read-zero conversion. The IN HI and IN LO terminals are shorted internally during a read zero. The reference selected by the other control bits is used. The read-zero conversion result is proportional to the internal offsets of the MAX133/134. This result should be subtracted from other measurements to obtain zero-corrected readings.

Filter On and Filter Short. These bits control the active filter (see Fig. 10-11).

DC. This bit selects the DC mode when set to 1 and selects AC when set to 0. The bit should also be set for ohms measurement.

External AC. This bit should be set to 1 whenever the AC mode is selected. Use 0 for DC.

Divider Sense. This bit, the 10-0 through 10-4, and the current bits select the input signal source. Divider sense should be 1 whenever the input attenuator is selected. Set divider sense to 0 to select the 400-mV input.

Current. Set divider sense to 0 and the current bit to 1 to select the current input. Although this bit and the associated pin are named "current," the actual input is the voltage drop across an external current-sensing resistor.

10.3 Component Selection

Observe the following notes when selecting components for the single-chip DMM.

10.3.1 Integration Resistors

For an accurate times-2 multiplication in the X2 mode, the two RINT resistors (see Fig. 10-2) must be equal. If the X2 mode is not needed, connect a 604-k RINT1 between BUF OUT 1 and the 4.7-nF integration capacitor. Leave BUF OUT 2 open. The value of both RINT1 and RINT2 is normally 301 k for a 545-mV or 655-mV reference. This sets the integrator output current to 2 μA during the deintegrate phase. Do not exceed 8-μA integrator current.

10.3.2 Integration Capacitor

The nominal value for the integration capacitor is 4.7 nF. This value, in combination with the integrator output current and the clock frequency, sets the integrator swing to about 3 V for the voltage ranges when RINT1 = RINT2 = 301 k, and the clock frequency is 32,768 Hz.

Although this same integrator swing can be obtained with other capacitor values by changing the value of RINT, lower values of CINT might introduce more noise through increased noise pickup and 50/60-Hz signals. Excessively high values of CINT also cause noise problems by reducing the integrator swing to unacceptably low values, causing the comparator noise to dominate the conversion errors. Large values of CINT also cause linearity errors because the settling time of the internal times-10 circuits is affected by the value of CINT.

The dielectric absorption of the integration capacitor directly affects the integral linearity. High-quality polypropylene capacitors are recommended. Polycarbonate and polystyrene capacitors might give satisfactory performance in less demanding applications. Polyester (Mylar) capacitors cause about 0.1% integral nonlinearity.

10.3.3 Active Filter

The RC time constant of the active-filter components shown in Fig. 10-6 sets the rolloff frequency. The effective value of RFILTER1 is the sum of its value plus the source impedance driving the filter. For example, at 30 V, the effective source impedance is the 101-k resistor in the attenuator. At 3 V, the effective source impedance is 1 M. This variable source impedance alters the filter characteristics somewhat as the different voltage ranges are selected.

The effect of the different source impedances can be minimized by increasing the value of the filter resistors and decreasing the filter-capacitor values in proportion. However, this increases the offset error caused by the A/D input-leakage current flowing through the filter resistors. For simplified design, use filter-resistor values between 1 M and 3 M.

A low rolloff frequency improves the normal mode rejection but at the expense of a longer settling time in response to input-voltage step changes. Aliasing is another consideration, particularly when an LCD bar graph is used with the DMM. If the bar graph is updated at 20 times per second and there is a 19-Hz component in the signal being measured, the beat frequency of 1 Hz appears on the LCD bar-graph display. To avoid aliasing effects, the filter time-constant is normally set less than 10 Hz. A 3-Hz rolloff (RC = 40 ms) further reduces the aliasing effects and increases normal mode rejection but still maintains an acceptable transient response with fast-varying signals.

As in the case of the integration capacitor, polypropylene capacitors are recommended. This is because dielectric absorption in the filter capacitors creates a small (but significant) time-constant settling error.

10.3.4 Crystal Oscillator

The oscillator is designed to use high-Q, low-power 32, 786-Hz crystals such as the Statek CX-IV. The series resistance should be less than 30 k. The oscillator capacitor connected to OSC2 (see Fig. 10-2) is typically 10 pF but should be adjusted to optimize performance with the chosen crystal. If overtone oscillations are observed, increase the value of the oscillator capacitor. Decrease (or eliminate) the capacitor if the oscillator has start-up problems. Keep stray capacitance across the crystal to a minimum. Such stray capacitance might prevent oscillation.

10.3.5 Attenuator Network

Figure 10-4 shows the attenuator network and the associated range-selection switches. If the resistance of the internal range-selection switches is 0 ohms, the theoretical values for the attenuator are 10 M, 1.1111 M, 101.101 k, 10.01 k, and 1.000 k. However, the important point to remember is that the resistors must track each other. This is because the ratio of resistor values sets the accuracy of the voltage measurements. As always, the temperature coefficients of the various attenuator resistors should be as low as practical. The voltage coefficient of the 10-M resistor should also be low because this resistor has high voltages applied (in the 400-V and 4,000-V ranges).

10.3.6 Attenuator Compensation Capacitors

The input attenuator is often compensated with low-value capacitors (see Fig. 10-2). These capacitors maintain a constant attenuation ratio over a wide bandwidth. Keep the capacitor values low to prevent the 10-M pin from being driven above V+ or below V− when high-frequency, high-voltage signals are applied. Such a condition can cause gross conversion errors.

10.3.7 PTC Resistor

As shown in Fig. 10-5, an adjustable PTC resistance is normally used as part of the protection circuit in the ohms mode. Excessive values of PTC resistance can reduce the voltage across the unknown and reference resistors, particularly at 400 ohms. PTC resistances greater than 2 k degrade system performance by reducing the signal level at 400 ohms, increasing the conversion noise. Values greater than 5 k cause additional error because the voltage drop across the PTC appears at the ADC as a common-mode difference between IN HI and IN LO.

10.3.8 Microprocessors

For low-cost two-chip digital multimeters, Maxim recommends 4-bit microprocessors with LCD display drive capability. Typical 4-bit microprocessor families include the Sharp SM4 and SM5, the NEC mPD75XX family, and the Hitachi LCD-III and LCD-IV families. If additional calculation power is needed, or if software development costs and time must be minimized, then 8-bit microcontrollers such as the 8048, 8051, or 6803 should be used.

10.4 Conversion Method

Figures 10-12, 10-13, and 10-14 show typical integrator and buffer waveforms for a large-positive, a large-negative, and a small-positive input voltage, respectively.

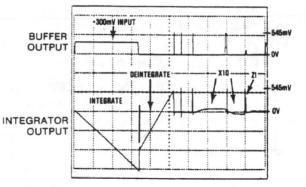

FIGURE 10-12
Waveforms for large-positive input (*Maxim Evaluation Kit Data Book,* 1994, p. 4-90)

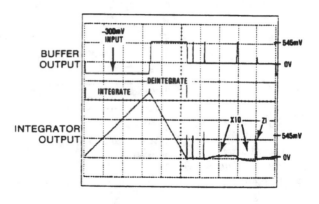

FIGURE 10-13
Waveforms for large-negative input (*Maxim Evaluation Kit Data Book,* 1994, p. 4-90)

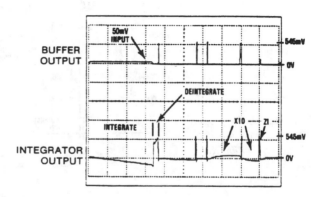

FIGURE 10-14
Waveforms for small-positive input (*Maxim Evaluation Kit Data Book,* 1994, p. 4-90)

The following is a summary of the "residue multiplication" technique used to perform a ±40,000-count conversion in 1,638 clock cycles.

10.4.1 Integration Phase

Figure 10-15 shows the integration periods for the various modes of operation. Figure 10-16 shows the basic elements of the ADC analog circuits. The unknown signal is integrated by connecting the noninverting input of the integrator to IN LO and the buffer input to IN HI. The integration period varies from 100 counts to 655 counts, as shown in Fig. 10-15. The IC is in the zero-integration phase while in hold (between conversions) and before the start of the integration period.

$$\text{Digital output code} = \text{Integration period} \times 100 \times V_{IN}/V_{REF}$$

where V_{IN} is the differential voltage applied to the ADC internal IN HI and IN HO and V_{REF} is the differential voltage applied to the ADC internal REF HI and REF LO shown in Fig. 10-16.

MODE	INTEGRATION PERIOD (clock cycles)	
Voltage, 60Hz	545	(16.63ms)
Voltage, 50Hz	655	(19.99ms)
Voltage, 60Hz, ÷ 5	109	
Voltage, 50Hz, ÷ 5	131	
Ohms	500	
Ohms, ÷ 5	100	

FIGURE 10-15 Integration periods for operating modes (*Maxim Evaluation Kit Data Book,* 1994, p. 4-90)

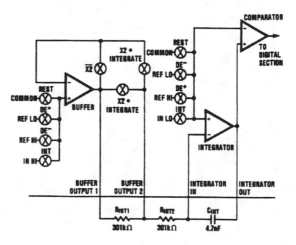

FIGURE 10-16 Basic ADC analog circuits (*Maxim Evaluation Kit Data Book,* 1994, p. 4-91)

10.4.2 First Deintegration Phase

The polarity of the first deintegration phase is determined by polarity of the voltage on the integration capacitor CINT at the end of the integration period. No reference capacitor is needed, improving the response time in ohms measurement. Also because the noninverting input of the integrator is connected to REF HI for a positive deintegration, the voltage at the integrator output has a step voltage-change equal to the reference voltage.

The first deintegration phase terminates when the comparator detects that CINT has been discharged. The IC then goes into an "idle" state in which both the buffer input and the noninverting input of the integrator are connected to common. This causes the system offset to be integrated. Near the end of the maximum allowable deintegration period, the polarity of the voltage on CINT is again tested, and either a positive or negative deintegration cycle occurs.

10.4.3 Times-10 (X10) Phase

When zero-crossing is detected at the end of a deintegration phase, the deintegration is continued until the next clock cycle. This causes the integrator to overshoot zero-crossing slightly, leaving a small residual voltage on CINT. Any comparator delay causes an additional residual voltage on CINT. The times-10 phase inverts and multiplies this residual by a factor of 10.

10.4.4 Second Deintegration Phase

The second deintegration phase deintegrates the residual voltage on CINT. (This voltage has been inverted and multiplied by 10 during the X10 phase.) Because the voltage across CINT is multiplied by 10, each clock cycle during second deintegration corresponds to $\frac{1}{10}$ of one clock cycle during the first deintegration phase.

10.4.5 Second X10 and Third Deintegration

The residual voltage on CINT after the second deintegration phase is multiplied by the second X10 phase, and this multiplied residual is deintegrated in the third deintegration phase. Because the residual voltage on CINT has been multiplied twice by 10, the third deintegration phase has 100 times finer resolution than does the first deintegration phase.

10.4.6 Sequence Counter and Results Counter

The sequencing or timing of the various conversion phases is controlled by a binary sequence counter. This counter counts upward continuously except during the hold mode. Some phases, such as the integration periods, are both started and stopped at present counts. The deintegration phases are started at predetermined counts but are terminated when the comparator detects zero-crossing at the integrator output.

The results counter accumulates counts during all deintegration phases. It is an up-down BCD counter, the count direction being determined by the deintegration polarity. The first deintegration phase causes the results counter to count by hundreds. Because the second deintegration phase is deintegrating a residual voltage that has been multiplied by 10, the results counter is incremented or decremented by ones during the third deintegration phase. The contents of the results counter are transferred to the results register at the end of each conversion.

10.5 Application Notes

The following manufacturer's recommendations should be considered when using the MAX133/134 for any application.

10.5.1 Sleep and Hold Modes

The hold mode stops the internal sequence counter at the end of the next conversion, but does not turn off the oscillator or any analog circuit. The hold mode can be used to accelerate autoranging (see Section 10.6.1). Dielectric absorption in CINT causes the first two or three readings after an extended hold period to have a lower magnitude than the steady-state reading.

The sleep mode puts the IC into a low-power quiescent mode by shuting off all analog circuits except the DGND power supply and the oscillator. A typical use for the sleep mode is to reduce power consumption by turning off the IC if the meter is idle for a long period. A typical method of detecting when the meter is no longer being used is to detect when the reading stays constant and there are no operator inputs (such as range or mode changes) for an extended period.

Because the sleep mode turns off all analog circuits, the first conversion after coming out of the sleep mode is not valid. It takes several readings before the reading has stabilized to within one count.

10.5.2 Input Protection for Digital Multimeters

Figure 10-5 shows a typical multimeter input circuit for ohms measurement. The adjustable PTC resistor (actually a thermistor) normally has a resistance of 2 k. However, under overload conditions, the PTC limits current because the current heats the PTC, drastically increasing the resistance.

Protection on the voltage ranges is automatic, because the 10-M input resistor limits the input current to safe limits, even with 4,000 V applied. The current ranges must be protected with fuses or circuit breakers. Current-sense resistors should be bypassed with diodes. As a simplified-design guide, voltage drop across the current-sense resistors should be limited to no more than two diode drops (typically about 1 V).

10.5.3 External AC-DC Converter

Figure 10-17 shows a half-wave external AC-DC converter. The circuit is an average-sensing, RMS-calibrated converter. This means that the output is proportional to the average AC value, rather than RMS value, but that the output is multiplied by 1.11 to correct for the ratio of average voltage to RMS voltage in a sine wave. If desired, a true RMS to DC converter can be used instead. Such a converter must be connected between the EXT AC IN and EXT AC OUT pins (in place of the Fig. 10-17 circuit).

10.5.4 PC Board Layout

Because the integrator output makes common-mode voltage steps equal to reference voltage (to perform a positive deintegration), any stray capacitance on CINT causes errors. Stray capacitive loading on the buffer output should be minimized to avoid ringing at the buffer output. The integrator-in node is particularly sensitive to stray pickup of noise and 50/60-Hz line noise, so keep CINT close to the INT IN pin.

Minimize capacitance on the node that joins the two RINT resistors. This capacitance sets up an RC time constant that rounds off the edges of the comparator input and can cause errors. If the times-2 mode is not used, connect a single RINT1 resistor directly from BUFF OUT 1 to the INT IN pin. Locate RINT1 as closely as possible to the INT IN pin (because the buffer output is a low-impedance point and INT IN is high impedance).

Any resistance between the 1-k pin and the 1-k resistor adds to the effective value of the 1-k resistor. This is also true of any voltage drop between the 1-k resistor and the IN LO pin. These resistances should be minimized or the 1-k resistor value should be reduced to compensate for the resistance of the PC board connections.

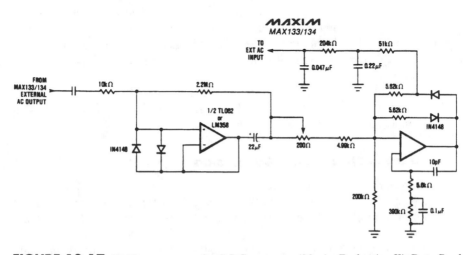

FIGURE 10-17 Half-wave external AC-DC converter (*Maxim Evaluation Kit Data Book,* 1994, p. 4-92)

The effective resistance of any current-sensing resistors is affected by where the voltage is sensed. Connect IN LO directly to one end of the current-sensing resistor to avoid errors caused by voltage drops in the common traces on the PC board.

10.6 Software Notes

The following manufacturer's recommendations should be considered when configuring the MAX133/134 and microprocessor software.

10.6.1 Autoranging

Figures 10-18 and 10-19 show the autoranging sequence when the IC is running continuously and when there is a hold between conversions, respectively. The sequence in which the registers are loaded has no effect, provided that all registers are loaded before the next EOC. Control bits take effect only when the IC is in hold or when the current conversion is completed.

If the IC runs continuously, the autoranging sequence is as shown in Fig. 10-18. If the IC is put into the hold mode during autoranging, the autoranging time can be reduced when several ranges must be tried. This is shown in Fig. 10-19.

A simple test that detects most overrange readings is to check if the two most significant bits (registers 3 and 4) are greater than ±45. A second test of the zero-corrected reading should be performed to make sure the reading is within the desired full-scale range.

10.6.2 Reducing Conversion Noise

The MAX133/134 has about ±1 counts of noise. In most cases, where only 4,000 counts are being displayed, averaging is not required because the noise is only $\frac{1}{10}$ of one displayed count. In data-acquisition systems (see Section 10.7) in which the full resolution is being used, averaging N readings reduces the noise by a factor equal to the square root of N. Thus if 50 readings are averaged, the noise is reduced by a factor of just over 7.

Because the noise of zero-corrected readings is the RMS sum of the noise of both the read-zero reading and the normal reading, the read-zero offset (see Section 10.2.5) should be averaged for best noise performance.

10.6.3 BCD to Binary Conversion

Normally, if only a zero correction or tare correction is to be applied to the IC output, the conversion result is left in BCD. If a scale factor or gain correction must be made, the result is usually converted to binary. Any of the standard BCD-to-binary conversion algorithms can be used.

A simple method of conversion is to read the MAX133/134 conversion result, starting with the most significant bit. Put the result into a multibyte accumulator and

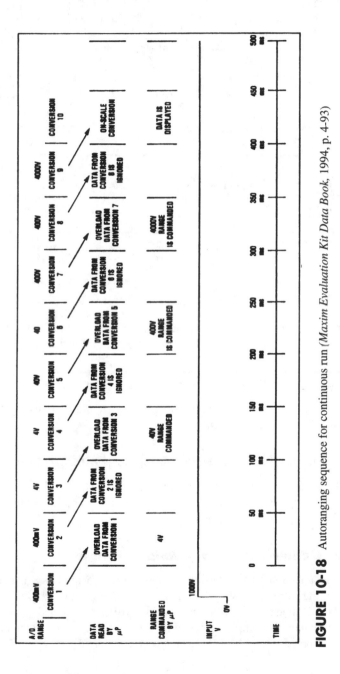

FIGURE 10-18 Autoranging sequence for continuous run (*Maxim Evaluation Kit Data Book*, 1994, p. 4-93)

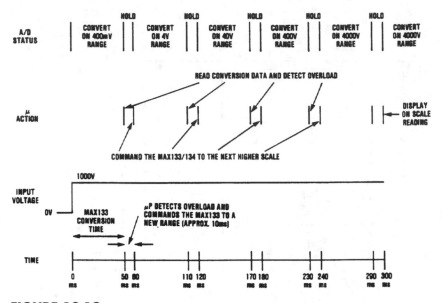

FIGURE 10-19 Autoranging sequence with hold between conversions (*Maxim Evaluation Kit Data Book,* 1994, p. 4-93)

multiply the result by 10. Then read the result for the next digit and add that to the accumulator. Repeat the "multiply-read-add" sequence for all five digits.

10.7 Using the IC in Data-Acquisition Systems

The following manufacturer's recommendations should be considered when using the MAX133/134 for data-acquisition systems.

10.7.1 Using the Input Attenuator as a Multiplexer

Figure 10-20 shows how the input section can be used as a multiplexer. This is suitable for data-acquisition systems in which the voltage range is limited, and the 400-mV to 4,000-V attenuators are not needed.

10.7.2 Using Nonstandard Voltage Ranges

In many data-acquisition systems, the voltage to be measured might have a full-scale range other than 400 mV, 4 V, and so on. For maximum resolution, the full-scale range should be adjusted to match the input-signal voltage span. This can be done either through attenuation or amplification of the signal (to make the signal match the basic ±400-mV span of the MAX133/134) or by means of adjusting the IC voltage span.

Figure 10-15 shows the integration periods of the various conversion modes. These different modes can be used to change the full-scale span of the IC. For exam-

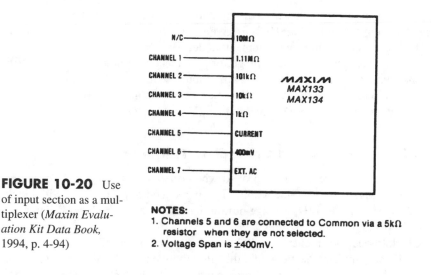

FIGURE 10-20 Use of input section as a multiplexer (*Maxim Evaluation Kit Data Book*, 1994, p. 4-94)

NOTES:
1. Channels 5 and 6 are connected to Common via a 5kΩ resistor when they are not selected.
2. Voltage Span is ±400mV.

ple, if the reference voltage is 545 mV, setting the 50-Hz bit changes the integration time to 655 clock cycles, and the 400-mV full-scale range becomes a 545/655 × 400 mV = 333-mV full-scale range.

Activating the ÷5 bit increases the full-scale span by a factor of 5. Setting the X2 bit decreases the full-scale span by a factor of 2 (assuming that RINT1 = RINT2).

In all cases, the values of RINT1, RINT2, and CINT should be chosen so that the integrator swing is at least 2 V, and integrator current is always less than 3 μA, during both deintegration and integration with a full-scale voltage. The common-mode voltage range of IN HI and IN LO is (V− +1.5 V) to (V+ −1.0 V).

10.7.3 Unipolar Operation

Unlike many integrating ADCs, the MAX133/134 does not have extra non-linearities when the reading is near zero. This makes it possible to use the full 80,000-count resolution to measure unipolar signals. All that is needed is a resistive offset network to translate the unipolar signal so that it becomes bipolar. An external zero circuit must be included so that errors in the offset resistor can be measured and subtracted. The zero-correction software is the same as would be used to correct for the internal zero error of the IC, except that in this case the external zero offset is nearly 40,000 counts.

10.7.4 Ratiometric Measurements

Ratiometric measurements (see Section 3.2.1) are used in many weigh-scale, pressure-transducer, and load-cell applications. If the reference voltage is referenced to ground or the common pin, simply connect the reference voltage to IN HI and IN LO, and perform any of the voltage-mode conversions.

When the reference voltage is a differential signal, use the circuit of Fig. 10-21. The programming table for Fig. 10-21 is given in Fig. 10-22. Use the bit pattern on Fig. 10-21 to select the ohms measurement mode.

The noninverting input of the integrator is connected to either REF LO or REF HI during deintegration. The integrator swing should be reduced if the integrator output goes within 0.5 V of either V+ or V−. In no case should either REF HI or REF LO be lower than (V− +1.5 V) or higher than (V+ −1.0 V).

10.7.5 Operation at Other Clock Frequencies

Noise performance can be improved when operating with clock frequencies lower than 32 kHz. However, the reading rate is reduced in proportion. With clock frequencies less than 10 kHz, leakage during the X10 phase introduces differential linearity errors at high temperatures.

Clock frequencies higher than 50 kHz are not recommended because the X10 period will not completely settle within the allotted time period, causing differential nonlinearity errors. Another potential problem at very high clock frequencies is that although the comparator delay is a fixed time period, the delay increases in terms of clock cycles as clock frequency increases. At very high clock frequencies, the residue cannot be fully deintegrated in the allotted number of clock cycles after having been multiplied by 10 in the X10 phase. When using a clock frequency other than 32,768 Hz, change the value of CINT to keep integrator swing at about 2 V.

10.7.6 Converting X2 to ±40-mV Full-Scale

The sensitivity of the X2 mode is increased by the factor

$$(RINT1 + RINT2)/RINT2$$

In the normal DMM application, RINT1 = RINT2 and the X2 mode increases the sensitivity of the IC by a factor of 2. If the two resistors have a 9:1 ratio, the X2 bit increases the sensitivity by a factor of 10. This can be used to get 1-μV resolution on a 40-mV scale.

10.7.7 Disabling the Active Filter

Because the source impedance in many data-acquisition systems is very low, the value of the filter resistors (RFILTER1 and RFILTER2) can be lowered to reduce the error caused by the ADC leakage current flowing through the resistors. If rapid settling is needed in a multichannel data-acquistion system, then the filter should be disabled. This is done by leaving the FILTER RESISTOR IN and FILTER RESIS-TOR OUT pins open and shorting the FILTER AMP OUT to FILTER AMP IN. However, do not leave the filter-amplifier connections open circuited. This can result in oscillation.

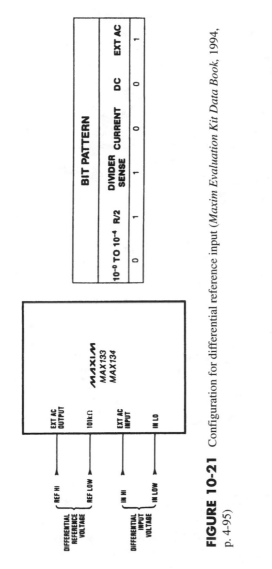

FIGURE 10-21 Configuration for differential reference input (*Maxim Evaluation Kit Data Book*, 1994, p. 4-95)

FIGURE 10-22 Programming table for differential reference input (*Maxim Evaluation Kit Data Book*, 1994, p. 4-94)

For Further Information

When applicable, the source for each circuit or table is included in the circuit or table title, so that the reader may contact the original source for further information. To this end, the complete mailing address and telephone number for each source are given in this section. When writing or calling, give complete information, including circuit title and description. Notice that all circuit diagrams and tables have been reproduced directly from the original source, without redrawing or resetting, by permission of the original publisher in each case.

AIE Magnetics
701 Murfreeboro Road
Nashville, TN 37210
(615) 244-9024

Dallas Semiconductor
4401 S. Beltwood Parkway
Dallas, TX 75244-3292
(214) 450-0400

EXAR Corporation
2222 Qume Drive
P.O. Box 49007
San Jose, CA 95161-9007
(408) 434-6400

GEC Plessey Semiconductors
Cheney Manor
Swindon, Wiltshire
United Kingdom SN2 2QW
0793 518411

Harris Semiconductor
P.O. Box 883
Melbourne, FL 32902-0883
(407) 724-7000
(800) 442-7747

Linear Technology Corporation
1630 McCarthy Boulevard
Milpitas, CA 95035-7487
(408) 432-1900
(800) 637-5545

Magnetics
Division of Spang and Company
900 East Butler
P.O. Box 391
Butler, PA 16003
(412) 282-8282

Maxim Integrated Products
120 San Gabriel Drive
Sunnyvale, CA 94086
(408) 737-7000
(800) 998-8800

Motorola, Inc.
Semiconductor Products Sector
Public Relations Department
5102 N. 56th Street
Phoenix, AZ 85018
(602) 952-3000

National Semiconductor Corporation
2900 Semiconductor Drive
P.O. Box 58090
Santa Clara, CA 95052-8090
(408) 721-5000
(800) 272-9959

Optical Electronics Inc.
P.O. Box 11140
Tucson, AZ 85734
(602) 889-8811

Philips Semiconductors
811 E. Arques Avenue
P.O. Box 3409
Sunnyvale, CA 94088-3409
(408) 991-2000

Raytheon Company
Semiconductor Division
350 Ellis Street
P.O. Box 7016
Mountain View, CA 94039-7016
(415) 968-9211
(800) 722-7074

Semtech Corporation
652 Mitchell Road
Newbury Park, CA 91320
(805) 498-2111

Siliconix Incorporated
2201 Laurelwood Road
Santa Clara, CA 95054
(408) 988-8000

Unitrode Corporation
8 Suburban Park Drive
Billerica, MA 01821
(508) 670-9086

Index